Ham Antenna Construction Projects

by

J. A. Stanley

WILDSIDE PRESS

Preface

Ham radio today is a competitive affair; at times the bands are jammed with stations, which means you can call repeatedly without making contact. This is *especially* true if your antenna is not adequate.

The quality of your antenna is undoubtedly the most important single element in determining how well you *get out.* The right kind of antenna can actually make a 50 watter sound like a "half gallon" half-way around the world; an inefficient antenna can prevent most of your RF from leaving the block you live in. Even worse, a poor antenna can add insult to injury by dumping most of your RF into nearby power lines, fences, and even into radios and TV sets—guaranteed to add nothing to your popularity in the neighborhood.

Just as is the case with ham transmitters and receivers today, excellent manufactured antennas are available. However, even if you already have manufactured units for several bands, you may want a couple of other antennas for use on additional bands.

Also, every antenna system is similar to a custom layout, because almost every home, apartment, and lot is different in size and shape. If you limit your antennas to those ready-made types you can buy, you may not be able to obtain the one you really should have to fit your particular need. By "rolling-your-own" you can solve such practical problems as getting real output on your favorite bands, while creating an antenna that can be put on top of the house.

Also, by building your own antennas, you can save an appreciable amount of money. In the process of building and tuning an antenna, you will learn how to extract the last watt of power from any antenna, homemade or manufactured. And finally, you will be able to experiment with various types of antennas before committing yourself to any one type.

All the antennas in this book are either homemade or represent major modifications of manufactured antenna elements. The designs are all practical, developed by hams from all over the world. Included are modern versions of such classic antennas as the Marconi, yagi, and 8JK. There are antennas of every size and description—antennas which require a city lot for erection, or which you can hang outside an apartment window or conceal in the attic. Among the selection is certain to be an antenna which will fit your particular layout and help you *get out*, even with low power.

The antennas described here are primarily intended for rigs in the 10- to 250-watt class. Some can be used, however, with more powerful rigs—in fact, any which do not use tuning capacitors as part of the hookup.

J. A. STANLEY

February, 1963

Contents

Chapter 6

Chapter 7

Chapter 8

Chapter 9

CHAPTER 1

Basic Horizontal Antennas

Radio transmission is accomplished by the fact that current traveling at a particular frequency along a wire cut to resonate at that frequency will efficiently radiate RF energy (Fig. 1-1). By observing the correct length of the wire used for the given frequency, the most efficient transfer of energy from transmit-

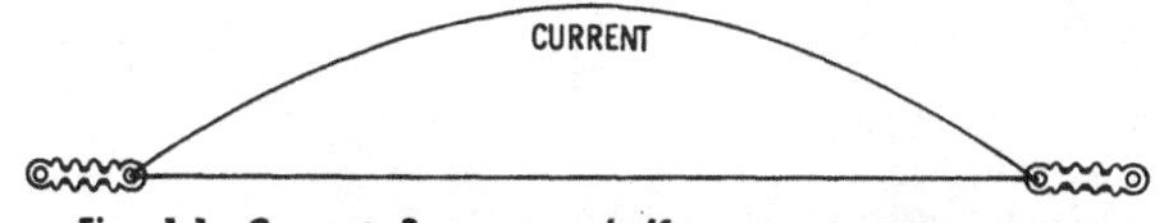

Fig. 1-1. Current flow on a half-wave resonant antenna.

ter to space is assured. In this manner, the antenna may be thought of as being tuned. Actually, this is no different from tuning an LC circuit to resonance.

Generally, the most widely used antenna is the half-wave; although there are many quarter- and eighth-wave antennas in use. The terms half-wave, quarter-wave, and eighth-wave are derived from the fact that a wire will resonate (although not too efficiently) at these, and other, multiples of the fundamental frequency. Consider, for example, a transmitter operating on a frequency of 30 mc (10 meters). A full-wave antenna for this frequency would measure approximately 32 feet in length. For a half-wave antenna, it would take only half as much wire (Table 1-1); for a quarter-wave antenna, only one-quarter as much; and so on.

DIRECTIONAL CHARACTERISTICS

Fig. 1-2 shows a cut-away view of the pattern of a horizontal antenna. Notice that the pattern is radiated in all directions perpendicular to the plane of the wire. With such a pattern, the antenna is said to be omnidirectional in the horizontal plane with little or no energy being transmitted or received from the ends of the wire.

Table 1-1. Half-wave Antenna Lengths (For Quarter-Wave, Cut in Half)

Frequency	Length	Frequency	Length
80 meter (3.5 mc)		**15 meter (21 mc)**	
3550 kc	131'-10"	21,050 kc	22'-3"
3600 kc	130'	21,100 kc	22'-2¼"
3650 kc	128'-2"	21,150 kc	22'-1½"
3700 kc	126'-4"	21,200 kc	22'-1"
3725 kc	125'-6"	22,250 kc	22'-½"
3750 kc	124'-8"	22,300 kc	21'-11½"
3800 kc	123'	22,350 kc	21'-11"
3850 kc	121'-6"	22,400 kc	21'-10½"
3900 kc	120'	22,450 kc	21'-10"
3950 kc	118'-6"	**10 meter (28 mc)**	
40 meter (7 mc)		28,600 kc	16'-5"
7050 kc	66'-4"	28,800 kc	16'-4"
7100 kc	65'-11"	29,000 kc	16'-3"
7175 kc	65'	29,200 kc	16'-2"
7225 kc	64'-9"	29,400 kc	16'-1"
7275 kc	64'-4"	29,600 kc	16'
20 meter (14 mc)			
14,050 kc	33'-4"		
14,100 kc	33'-2½"		
14,150 kc	33'-1"		
14,200 kc	32'-11½"		
14,250 kc	32'-10"		
14,300 kc	32'-8"		
14,350 kc	32'-7"		

Looking at the antenna from the top, as in Fig. 1-3, it can be seen that, with a wire strung from East to West, the radiation pattern is in a North-South direction. Note that not only is there a strong radiation pattern from North to South (*A* in Fig. 1-3), but there are also strong patterns from Northeast to Southwest and from Northwest to Southeast (*B* and *E*). This, of course, allows maximum radio coverage over a large area of the earth.

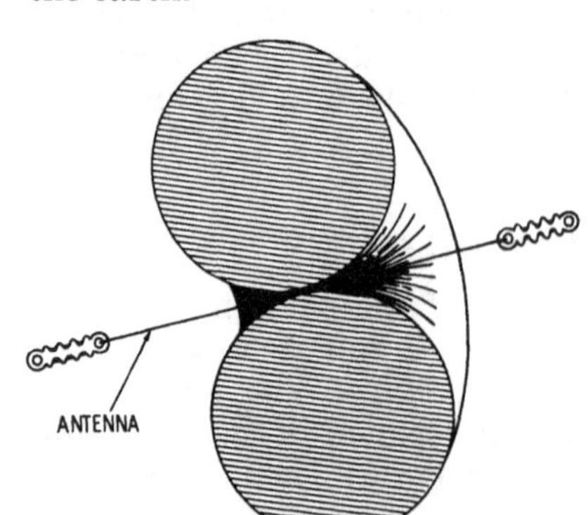

Fig. 1-2. Cross-sectional view of the radiation pattern of a half-wave antenna.

ANTENNA FEED-POINT IMPEDANCE

An antenna simply hanging up in free space isn't worth much; we need to get our energy from the transmitter up *to* the antenna. Of course, we can bend down one end of the antenna and bring it into the house—but, except for certain special types, this is not the best idea for several reasons. Some

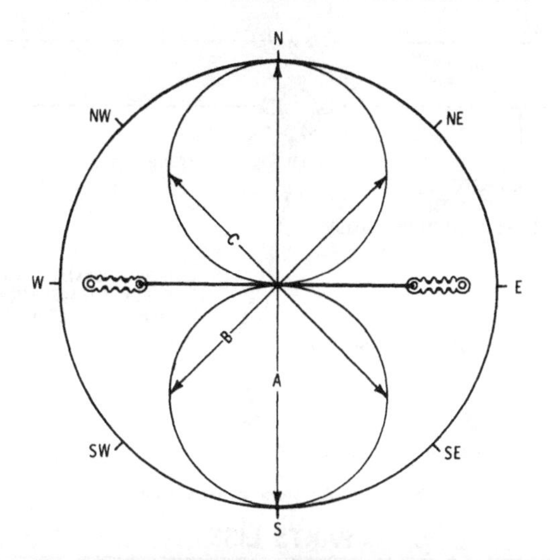

Fig. 1-3. Top view of a horizontal-antenna pattern.

energy may be wasted because of absorption by nearby objects. Also, the normal radiation pattern of the antenna may be somewhat upset.

For these reasons, in most cases, some type of feeder line is used to get the energy from the transmitter to the antenna (Fig. 1-4). At this point we need to consider another matter: the impedance of the antenna at the point where the feeder must connect. Fortunately, the theoretical input impedance of a half-wave antenna fed in the center is a value which can be matched by a feedline which, in turn, is of a handy impedance value for powering from the common ham transmitter. Such an arrangement is the basis of the first antenna we will consider.

THE COAXIAL-FED, ONE-BAND DIPOLE

In many ways this type of antenna is the simplest to build because it can be used without any tuning devices. In addition, the mechanical construction is quite simple.

The first step in building the coaxial-fed, one-band dipole is to determine space requirements for such a device. Table 1-1 gives the dimensions for antennas in each of the key bands. Select the antenna length which is closest to the frequency you plan to use the most. This will show you how much wire you need. The parts needed will be:

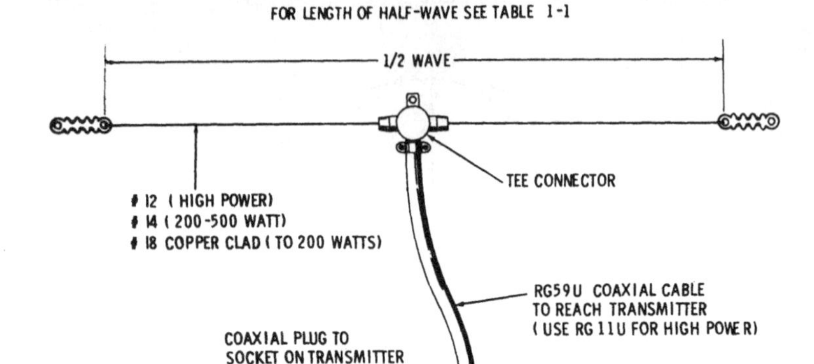

Fig. 1-4. Antenna construction.

PARTS LIST

Quantity	Description
1	length of antenna wire. #18 Copperweld steel okay for use up to 200 watts. #14 or #12 hard-drawn copper (or Copperweld) for higher powers.
1	length of RG-59/U of proper length for half wave. See Table 1-1.
2	seven-inch, high-quality, strain-type insulators (or use several smaller insulators in series).
1	coaxial plug and tee connector.

The easiest way to build this type of antenna is to stake it out on the lawn as shown in Fig. 1-5. Cut the wire to the proper length as indicated in Table 1-1 (to the frequency you will likely use most). Notice that the figure given in the Table is for the overall length—each section of the antenna should be one half that size. Also, allow 3 inches of wire on each end to provide wire for hooking on an insulator.

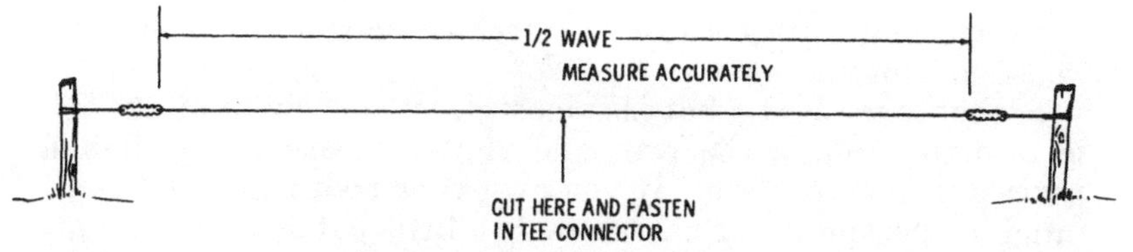

Fig. 1-5. Antenna being laid out.

Now cut the wire exactly in the middle and insert the dipole coupler. Hooking up this connector will vary in method depending upon the brand of coupler used; Fig. 1-6 shows how to connect up one popular unit. Be certain that the clamp on the connector clamps down hard on the cable cover. This is important to avoid excessive strain on the connections. There may be a bushing provided to help insure effective clamping action on RG-59/U, which is smaller than RG-8/U for which the connector is designed to handle.

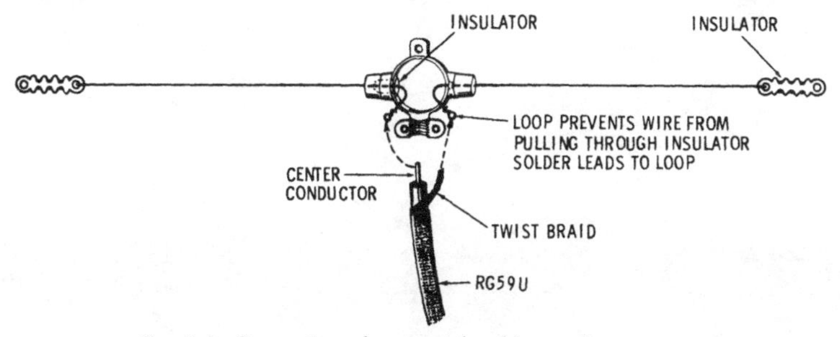

Fig. 1-6. Connecting the coaxial cable to the antenna.

Masts

The next step in building the antenna is to get it up in the air. There are many ways of doing this; Fig. 1-7 shows a couple of suggestions. Also, you will want to carefully consider the possibility of using some type of TV antenna mast. There is almost every type imaginable available, and because they are mass produced, they are good values. For example, the telescoping type of masts, properly guyed, work fine in many installations. If you have space on top of your house or garage, a 10-foot mast, properly guyed and mounted on some convenient type of TV-antenna mount, provides a good way to do the job.

Getting the Feeder Inside

Once you have the cable connected to the antenna, the next step is to get it into the house. At this point you'll have to use some ingenuity because the problem varies so much from house to house.

Probably the best—but also most drastic—way to do the job is to drill through the wall and run the cable in through a porcelain-tube insulator. If you own your own home and don't mind the possibility of having to do a little patching and painting on the inside, this approach is practical. If you use a TV-type wall feed-through insulator, chances are good that the plastic end flanges will cover up any small nicks around the hole you put through the wall.

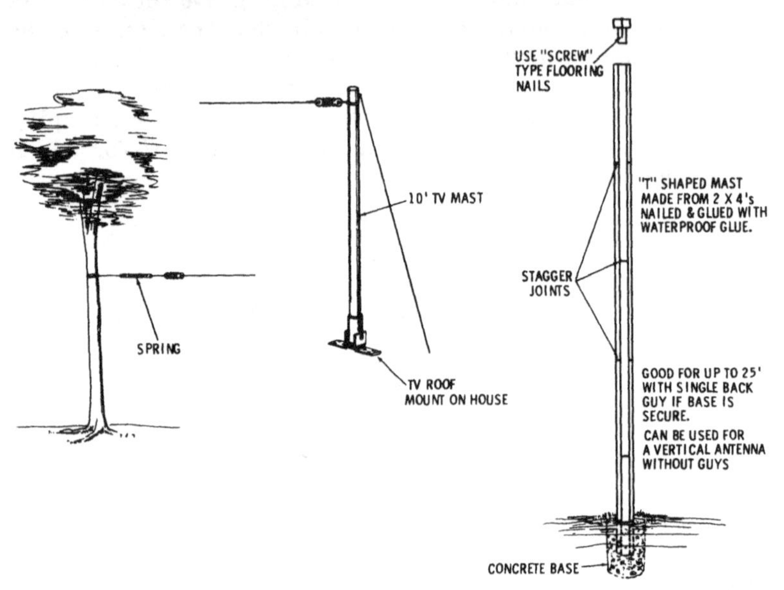

Fig. 1-7. Supporting the antenna.

Less drastic is the use of some type of board under a window, and then use a porcelain tube in the board. Yet another approach is to drill the glass. However, this is a job best done by a professional glazier. In the case of windows with small panes, one pane can be removed and replaced with clear plastic, which is easy to drill. Regardless of which of the methods you use, remember that what you want is a good waterproof job, and one which is mechanically solid so that the feed line doesn't whip about and knock out a window.

Once you have the antenna feed line inside the house, the final step is to hook it into the transmitter. It is assumed that you have a typical, modern rig with a Pi-Network output circuit intended for coupling into a low-impedance line. If so, the

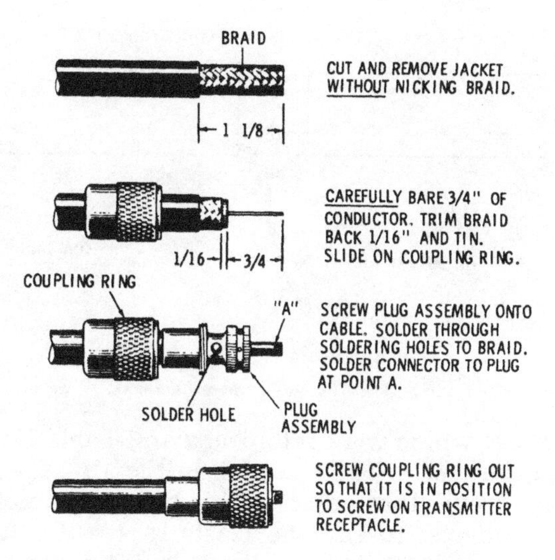

Fig. 1-8. Soldering the connector to the cable.

feed line should be fitted with a plug which matches the terminal on the transmitter (or change-over switch, if you have one between the antenna and transmitter).

Fig. 1-8 shows how to solder to the most common type of plug. Do the job *carefully*. If you make a mistake and accidentally short the center conductor to the outside braid, you will completely wreck any chance of the antenna working properly. Once you have the plug on the cable, you only have to plug it into the transmitter and follow the tuning instructions which came with the transmitter.

FOR THE NOVICE; 40-15 METER ANTENNA

Because these two bands happen to be in the proper kind of harmonic relationship, a coaxial-fed dipole (doublet) cut for 40 meters (Fig. 1-9) will also work without change on the 15-meter band. However, it is *not* suitable for other bands.

Improving the Doublet

It is a pretty safe bet that most of the doublets in use today are simply cut to measurement as described, hoisted, and forgotten. Most of them work well enough when built in this fashion. However, you may want to go one step further and endeavor to make your antenna a bit more efficient by making certain it actually *is* resonant at the frequency you have chosen. The reason that it might not be resonant is that all the calcula-

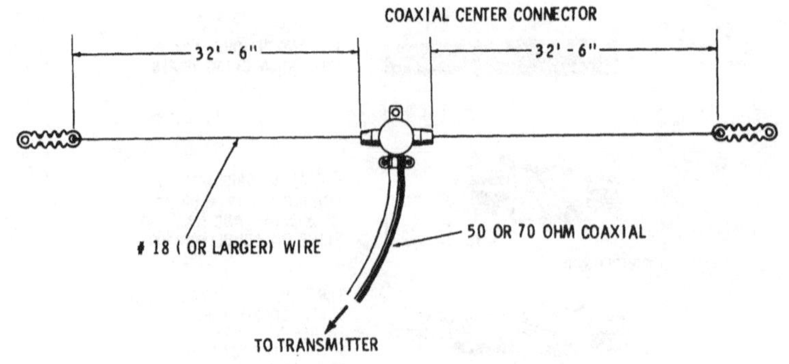

Fig. 1-9. The 40-15 meter antenna.

tions like those which went into preparing Table 1-1 are based on the idea that the antenna is high in the air—free from any interference from other objects. This happy state of affairs is difficult to achieve, to say the least. In reality, it is pretty common for an antenna to be a bit off the ideal resonant length.

One way to do the job is to deliberately cut the antenna a bit short, and then piece out both ends with some perforated aluminum strips, as shown in Fig. 1-10. Then, with a field-strength meter (Chapter 3), the antenna can be adjusted for the best length.

This is not an easy procedure. It is difficult to maintain the same output all of the time from the transmitter. Ideally, the transmitter should be tuned up for a light antenna load, and then tuning adjustments at the transmitter left alone. Also, keep your test periods short to avoid interfering with other stations. Alternately, the job can be done with an SWR meter, adjusting the antenna to give minimum SWR at the desired frequency (Chapter 3).

FOLDED DIPOLE

Another way to help insure that the antenna is close enough to resonance for maximum efficiency is to substitute a folded dipole for the single wire in the "flat top" (the horizontal portion of the antenna). Such an antenna is less critical to fre-

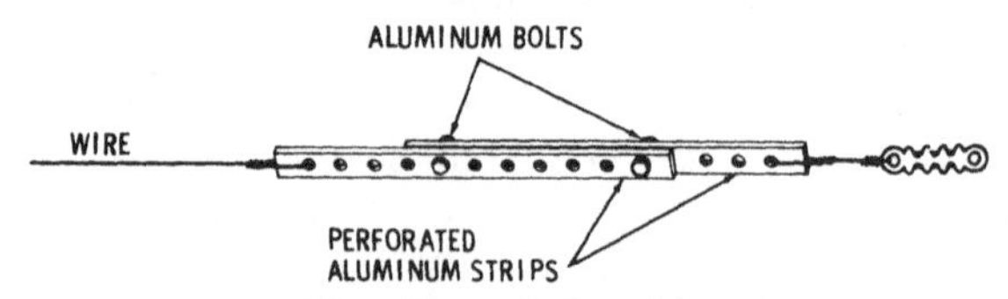

Fig. 1-10. Adjustable end piece for antennas.

quency. In fact, a folded dipole out to the *center* of an amateur band ordinarily will be broad tuning enough to give good results completely across the band.

An ordinary single-wire dipole would be far more frequency sensitive; and in the case of an amateur band like 80 meters, if the dipole is cut for the 3,500-kc end of the band, it will not work well in the 3,900-kc region. If you want good results from one end of the band to the other, the folded dipole is definitely one good answer. Also, the folded dipole is a truly balanced system which keeps stray radiation from the feed line at a minimum. The antenna does create a bit of a problem, though, in that it presents a balanced 300-ohm impedance which, with most transmitters, must be matched to an unbalanced 72-ohm feed point. One very practical way to do this is to use a device called balun coils. These coils not only match from the balanced to the unbalanced line, but also reduce the impedance at a 4 to 1 ratio (from 300 ohms down to 72 ohms).

Building a Balun-Coil Coupler

Fig. 1-11 shows a complete balun coil setup. Such a unit can be easily built using parts made by at least two different manufacturers. Don't despair if your parts jobber doesn't stock the coils; in most cases he can order them for you. If not, one of the mail-order jobbers can do it.

PARTS LIST

Quantity	Description
1	2 × 7 × 7 chassis.
2	feed-through insulators.
1	pair of Balun coils (B & W type 3976).
1	SO-239 coax receptacle.
	miscellaneous 6-32 bolts and nuts.

Important: a metal chassis, or some other type of metal base, is necessary for proper operation of balun coils.

In building the unit, first of all drill the holes and mount the parts. (Be careful in handling the coils, which are rather easily bent.) Then, following Figs. 1-12 and 1-13, do the wiring job (the unit shown was wired with bare wire, but this is by no means necessary). Check the wiring carefully. The wiring job is simple, but it is the kind of unorthodox circuit which breeds wiring errors. Also, before starting, check the coil instructions against the diagram to see if the terminal arrangement is the same as shown.

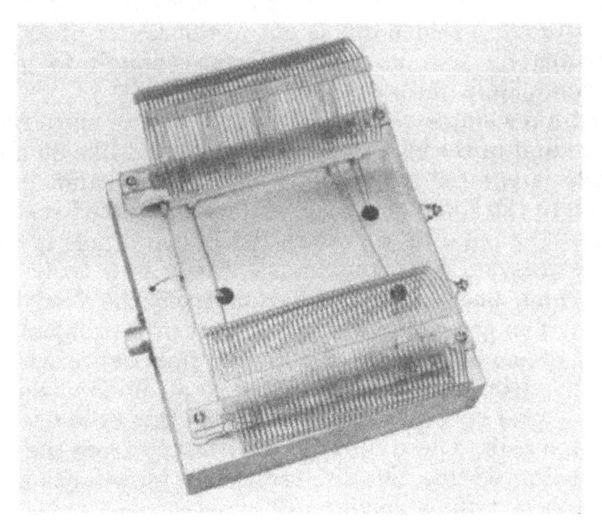

Fig. 1-11. A balun-coil coupler.

Balun Coils ready? Let's look at the overall hookup of a folded dipole as shown in Fig. 1-14. Note that the theoretical figure for overall length is 462/f mc, which is somewhat shorter than the usual 468/f mc. Calculate for the frequency

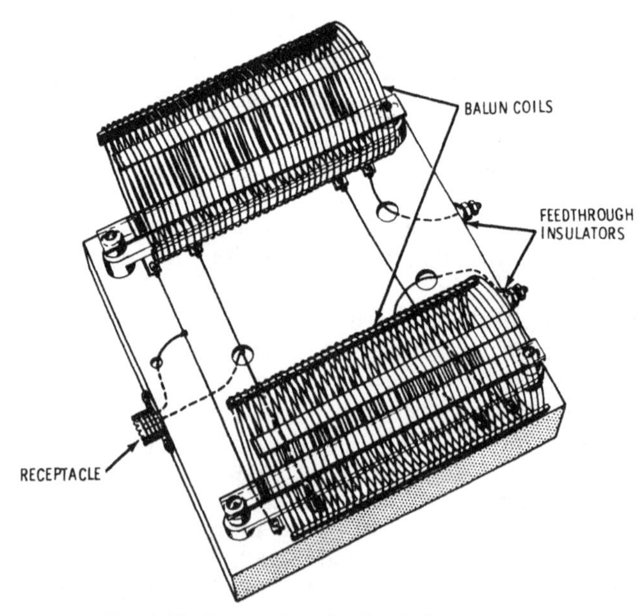

Fig. 1-12. Construction details of the coupler.

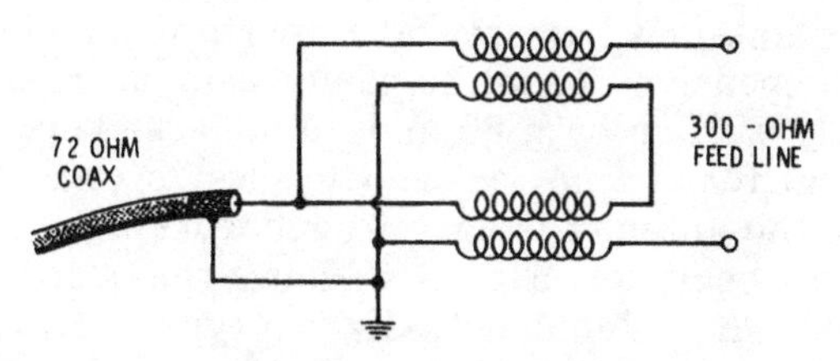

Fig. 1-13. Schematic of the coupler.

you want, or use Table 1-2. This table gives the length of a folded dipole cut for the center of each amateur band. Once the hook-up shown in Fig. 1-14 is complete, no other tuning is necessary. The balun coils do the matching job from 80-10 meters.

Table 1-2. Folded-Dipole Antenna Lengths

Band	Length
80 Meters	123'-6"
40 Meters	64'-6"
20 Meters	33'-6"
15 Meters	21'-9"
10 Meters	16'-0"

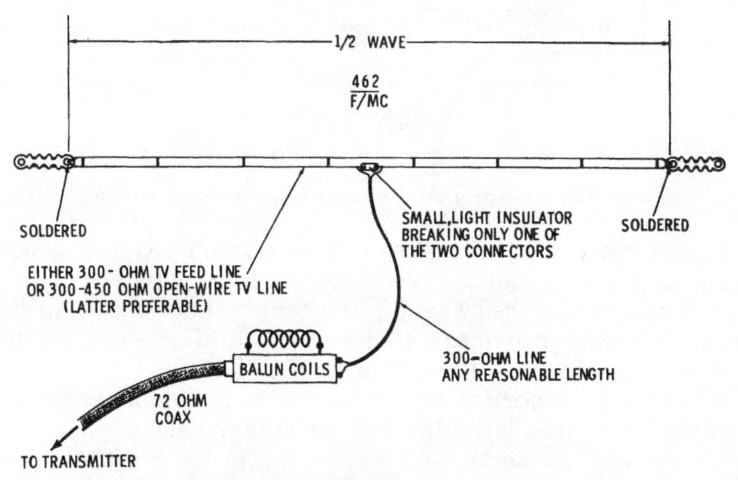

Fig. 1-14. Coupler inserted between transmitter and antenna.

THE NOVICE-90

The folded dipole is one way to overcome the problem of having an antenna work over only a relatively narrow range of frequencies. Another solution is to use an antenna which is tunable from within the room in which your ham rig is operating. Such an antenna has several important advantages. Not only is it inexpensive, it can be easily kept on resonance and works equally well on both 80 and 40 meters. This, of course, makes it ideal for the novice who wishes to operate on both bands. Also, the antenna is easy to get working, thus making it ideal for the beginner. Fig. 1-15 shows the overall layout of the Novice-90. The antenna actually is a type of Marconi—a 3/8 wavelength antenna on 8 meters and a 3/4 wavelength antenna on 40 meters. The extra length over the standard 1/4 wave

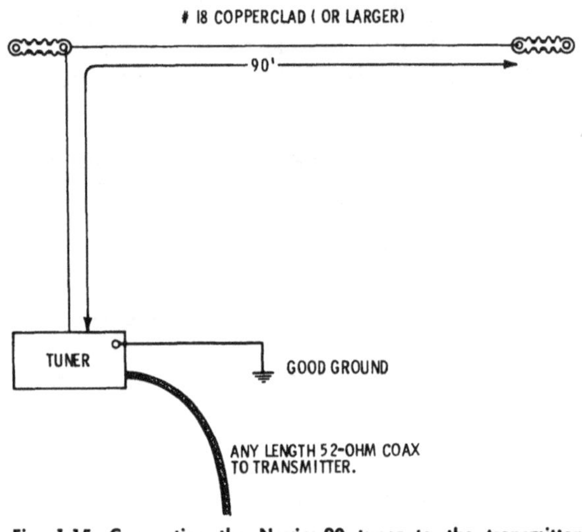

Fig. 1-15. Connecting the Novice-90 tuner to the transmitter.

Marconi raises radiation resistance (which reduces ground loss) and also permits a good match to a standard 52-ohm line. The heart of the system is the simple tuner shown in Fig. 1-16. This tuner provides a choice of two tuning arrangements, the circuits of which are shown in Fig. 1-17.

The radiating portion of the Novice-90 is simplicity itself—a 90-foot L-shaped wire. Ideally, the antenna should be erected with the vertical portion 30-feet long and the horizontal portion 60-feet long, but the proportions are not at all critical. One end of the wire should be brought into the operating room to connect to the tuner (which should be mounted close to the window or other point of entry). Since there is only a single wire, if necessary, you can simply wrap it with 10 turns (or more) of high-voltage electrician's tape and close the window on it. The tuner, as shown in Fig. 1-15, simply connects to the transmitter by means of a 52-ohm coaxial line.

Fig. 1-16. The tuner used with the Novice-90.

Obtaining a Good Ground

With any type of grounded antenna, a low-resistance ground is most important. As mentioned previously, the Novice-90 has the advantage of having fairly high radiation resistance. Therefore, the ground is not as critical as with some other types of antennas. However, attention paid to obtaining an efficient ground will definitely improve antenna performance.

One good way to do this is to provide a chemical ground. This involves driving a grounding rod into the earth and then filling a small trench (or a series of holes) around the rod with

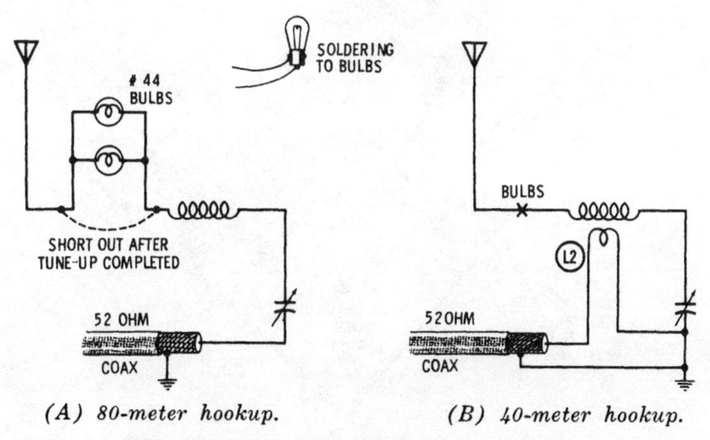

(A) 80-meter hookup. *(B) 40-meter hookup.*

Fig. 1-17. Circuits of the Novice-90 antenna tuner.

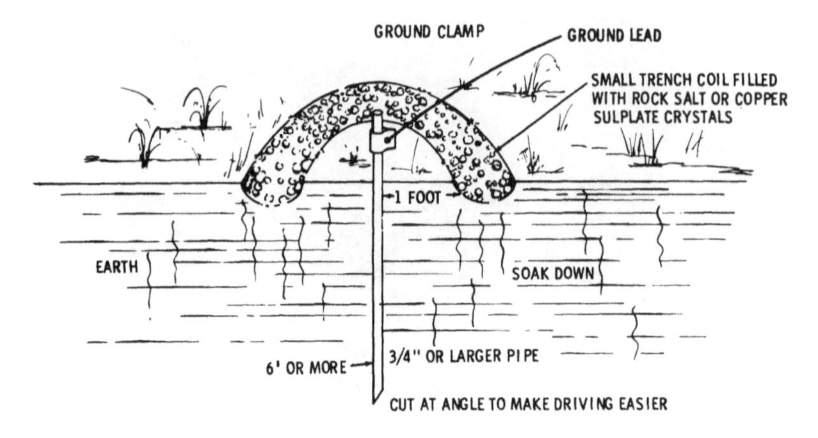

Fig. 1-18. A chemical ground for the Novice-90.

a conducting chemical, such as rock salt, or copper sulphate crystals (Fig. 1-18). An ideal place for such a ground is in a basement window-well alongside the house. (A window-well gathers moisture naturally.)

Building the Tuner

Mount the tuner parts as shown in Figs. 1-16 and 1-19. Fig. 1-17 illustrates the circuit diagram. You should check wiring

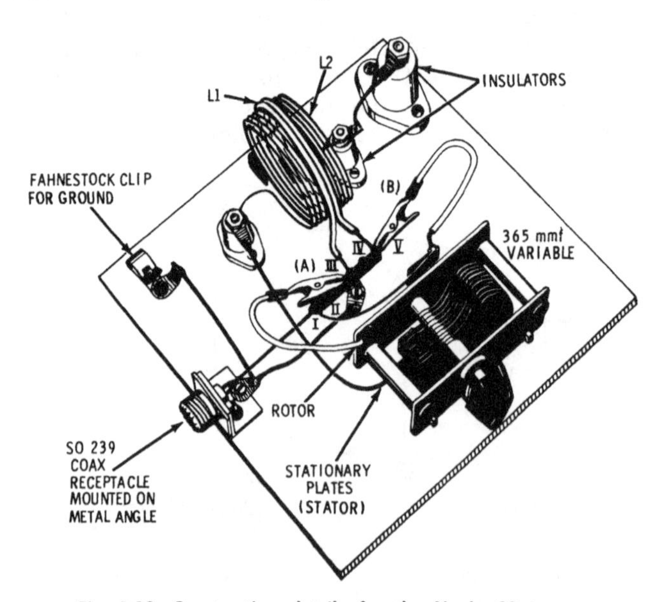

Fig. 1-19. Construction details for the Novice-90 tuner.

against both pictorial and schematic diagrams to make certain there are no wiring errors.

PARTS LIST

Quantity	Description
1	8 × 8 baseboard.
1	365 mmf (exact size not critical) receiving-type variable capacitor with tuning knob.
1	12 turns of 2″ inside-diameter air-wound coil. #16 wire (B & W 3907-1, or see text).
1	SO-239 coax-cable receptacle.
2	terminal strips.
3	stand-off insulators.
2	alligator clips.
1	short length of high-voltage picture-tube cable.
12	wood screws.
1	2½′ length of bare copper wire for hookup.
2	short lengths of flexible stranded wire.
1	metal angle.

Making Coils

For the coils you can use the commercial coil stock specified, or make your own air-wound coils. Fig. 1-20 shows a finished coil ready to be cut apart after pulling out the paper core. The coil is made by the following steps:

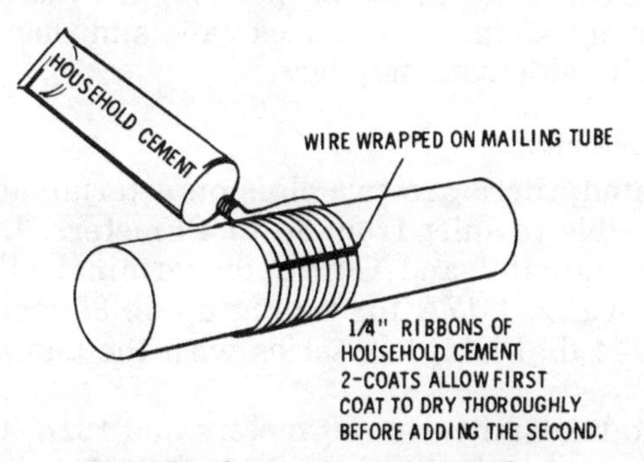

Fig. 1-20. Constructing air-core coils.

1. Wrap a mailing tube of the proper diameter with several layers of wax paper.
2. Take the kinks out of the wire by pulling it through a cloth held tightly in your hand.

3. Secure one end of the wire to a vise or other firm support (Fig. 1-21).
4. Secure the other end of the wire to the mailing tube.
5. Rotate the tube, rolling the wire onto it. Keep tension on the wire as you rotate the tube.
6. Secure the wire to the tube by threading it through a couple of holes.
7. Spread a quarter-inch ribbon of Dupont household cement on the coil in 4 places.

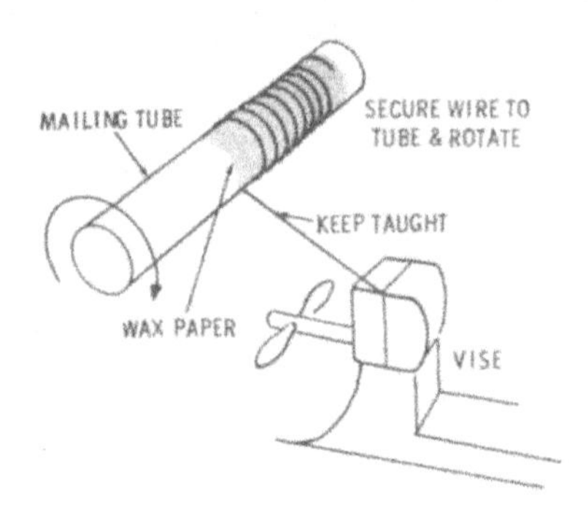

Fig. 1-21. Winding the coil.

8. Allow coil to dry thoroughly. Place a second layer of cement on top of the first. Again allow to dry thoroughly.
9. Once the glue dries, pull the mailing tube apart with side-cutting pliers. This will leave the coil, supported by strips of cement, now dried to plastic-like hardness.
10. Cut the coil stock to the proper length by sawing through the cement strips with a hack saw, and then clipping the wire with side-cutting pliers.

Tuning to 80 Meters

As indicated, there are two clips on a terminal strip which make it possible to shift from 80 to 40 meters. Initially, snap clip *A* on terminal *I* and Clip *B* on terminal *IV*. This gives circuit *AA* in Fig. 1-17A for tuning up on 80 meters. Connect paralleled #44 dial lamps in series with the antenna lead.

1. Set the transmitter to 80 meters and turn it on.
2. Adjust the antenna load capacitor to about one-half capacity.
3. Dip the transmitter plate-current meter for minimum (or in the case of some transmitters, adjust for maximum output meter).
4. Adjust the variable capacitor on the tuner for maximum

output as indicated by the glow of the bulbs. At this point, the antenna is tuned to resonance.

5. Now readjust the antenna load capacitor on the transmitter to achieve the desired load. As a final step, recheck the antenna resonance with the bulbs, making whatever capacitor adjustments necessary to achieve resonance. (The reason for using two bulbs is to avoid burning them out. If the bulbs glow only dimly, one bulb is enough, and will give a better indication.) Whether you are using one bulb or two, remove it from the hook-up after the tuning is completed.

Tuning to 40 Meters

On 40 meters, the tune-up procedure is the same as on 80 meters. However, clip *A* should be moved to terminal *II* and clip *B* to terminal *V* (Fig. 1-19). Then proceed exactly as for tuning up on 80 meters.

With most transmitters, pick-up loop L2 will provide plenty of coupling to allow loading the transmitter to the maximum desired plate current. If, even with the tuner on resonance as indicated by the bulb or bulbs in series with the antenna, there is not sufficient loading, add a turn or two to coil L2. *Important:* Always remove (or short across) the tuning bulb after the tuning process is completed.

FIVE-BAND TUNABLE ANTENNA

In recent years, antennas with a number of parallel tuned traps have seen a lot of use largely because they require no tuning in changing bands. This convenience is achieved with several sacrifices of efficiency:

1. The resonant frequency range on each band is usually limited, particularly on 80 and 40 meters.
2. On the higher frequencies (for example 28 mc), most of the antenna is unused, thus failing to take advantage of the low-angle and beam effect achieved when a horizontal antenna is operated on harmonics.
3. Unless a tuned circuit of some type is inserted in the antenna feed line, there is the strong possibility of radiating harmonics which can get you into trouble with the FCC.

One practical solution to all of these problems is an antenna which combines the features of the Novice-90 and the time-proven Zepp. On 80 meters, the antenna is a 3/8 wave Marconi just like the Novice-90. On 40 meters it becomes a ¾ wave antenna, end-fed. On 20 meters, it is a full-wave antenna fed this time with tuned feeders. On 15 meters—1.5 wavelength. On 10 meters, it is 2 wavelengths long (a length which gives 1.5-db gain, the equivalent to raising power 50%), and more important, it concentrates radiation at the low angles necessary for DX.

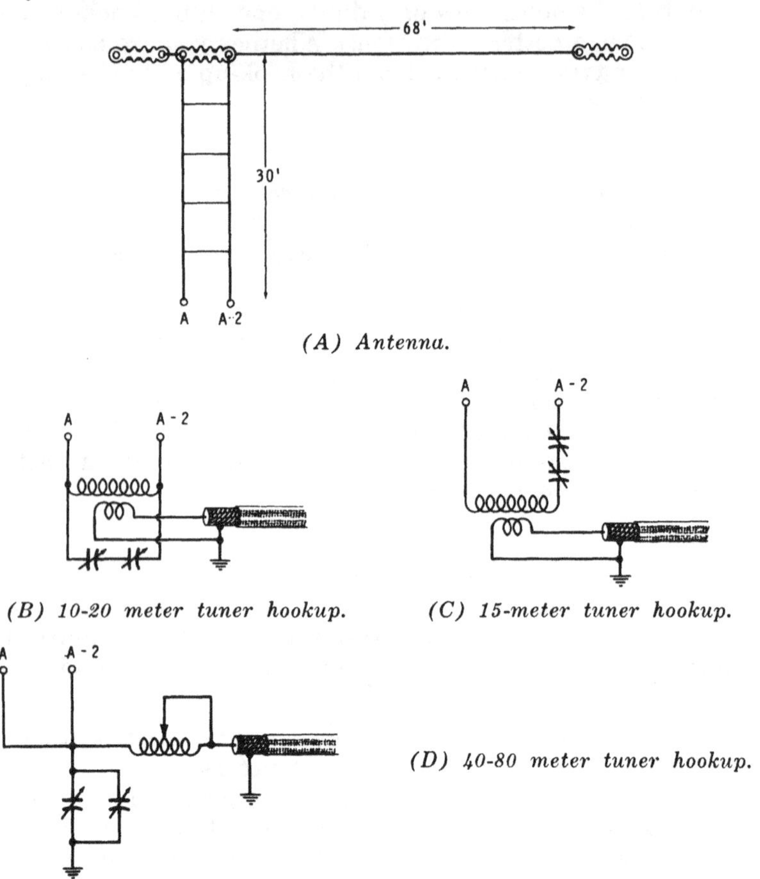

(A) Antenna.

(B) 10-20 meter tuner hookup.

(C) 15-meter tuner hookup.

(D) 40-80 meter tuner hookup.

Fig. 1-22. 5-band tunable antenna.

Fig. 1-22 shows the circuit layouts for the antenna on the different bands. Notice that three different coil hook-ups are needed. This could be accomplished with an elaborate coil-switching system, but a simpler way is to make up a plug-in coil for various bands. By arranging the connection to the coil plugs in the proper fashion, the desired arrangement is auto-

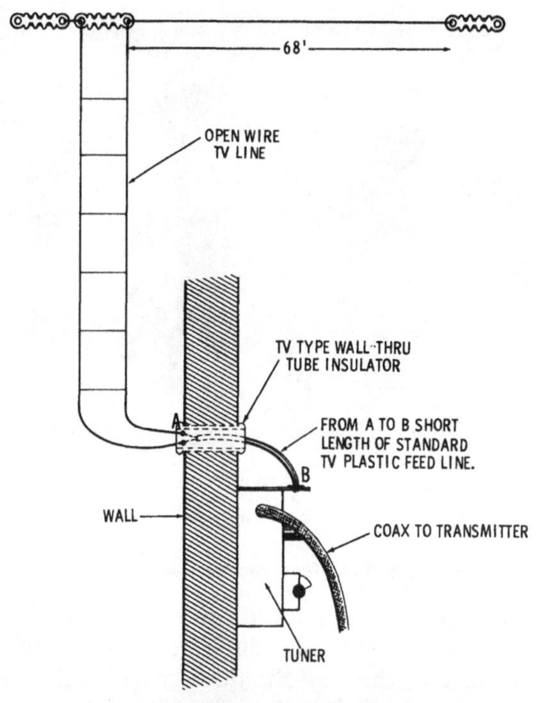

Fig. 1-23. Antenna-construction details.

matically achieved simply by plugging in the coil for the band to be used.

Erecting the Antenna

Fig. 1-23 shows the overall layout of the antenna with the various parts specified. Note the open-wire TV-feeder line—ideal for power up to 250 watts. Using it will save you the work of making an open-wire line. Further, the feed line can be used with the special stand-off insulators available for such line, thus simplifying installation work. Also note that one side of the feeder simply runs up to an insulator and stops; the other feeder actually doing the job of powering the antenna.

A TV-type lead-in tube is an ideal device for terminating the open-wire line. A short length of standard 300-ohm plastic line will serve the purpose of connecting the open-wire line to the tuner inside the house.

Building the Tuner

Fig. 1-22 gives the circuit of the tuner, and Figs. 1-24 and 1-25 show the layout and hook-up. Note that the dual 75-mmf

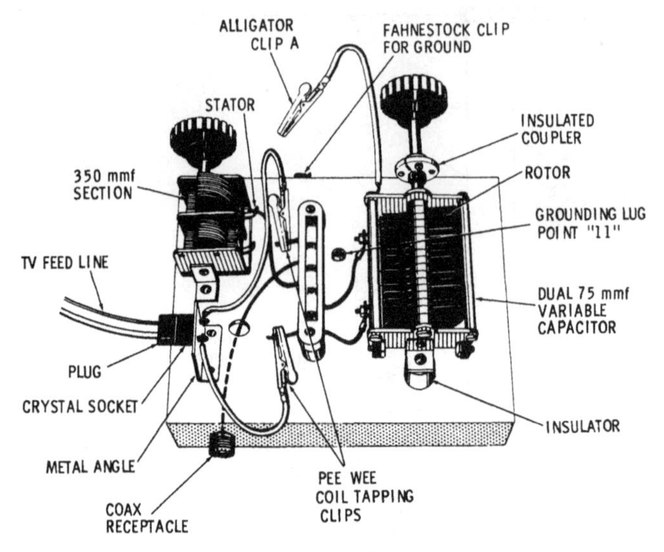

Fig. 1-24. Tuner-construction details.

variable capacitor is mounted on insulators. The smaller variable can be mounted on the chassis directly. As in any device used for RF, this one should be wired with heavy wire—#14 bare copper wire is ideal.

10-Meter Coil—This coil is made up of 5 turns of #14 wire, spaced to occupy approximately 8 turns per inch. BYW #3900 coil stock is suitable, or you can roll your own, as previously described.

The coil is first mounted on the plug bar, with the ends run through the proper pins. Fig. 1-26 shows how to make con-

Fig. 1-25. The completed tuner.

PARTS LIST

Quantity	Description
1	5-prong Steatite jack bar (Millen 41305).
4	5-prong Steatite coil plug (Millen 40305).
1	dual 75-mmf transmitting-type variable capacitor.
1	TV-type coil.
2	⅝″-cone Steatite cone insulators.
2	1″-cone Steatite cone insulators.
1	SO-239 (831R) coaxial chassis receptacle.
1	crystal socket for antenna feed-line plug.
2	thin-nose phosphor-bronze pee-wee coil clips (Mueller #88).
1	copper mini-gator clip (Mueller #30C).
1	insulated shaft coupler.
1	length #3900 BYW coil stock.
1	length #14 base copper wire (for hook-up wire).
1	superhet-type variable capacitor (larger section 350 mmf or larger).
2	knobs.
1	bracket for mounting crystal socket.
1	2 × 7 × 7 aluminum chassis.
1	fahnestock clip for ground.
	miscellaneous 6-32 machine screws and nuts.

nections. Next, the pick-up link is wound on (over the coil) and connected to pins 3 and 4. Finally, pin 2 is wired to pin 5 to complete the circuit.

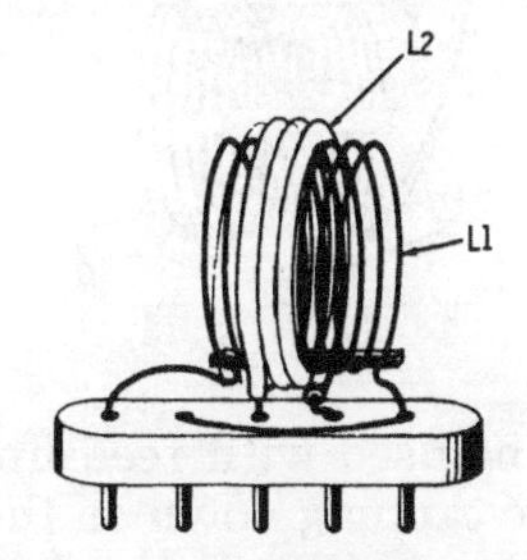

Fig. 1-26. Coil-pin hookup for 10-20 meter coils.

20-Meter Coil—The overall method for constructing this coil is the same as that for the 10-meter coil. However, this time, the main tuning coil has 10 turns, and the pick up loop 3 turns.

15-Meter Coil—Here too, the method of construction is the same. However, a different hook-up is used to provide series tuning, as shown in Fig. 1-27.

80-40-Meter Coil—For the construction of these coils see Figs. 1-25 and 1-28. This hook-up transforms the circuit into

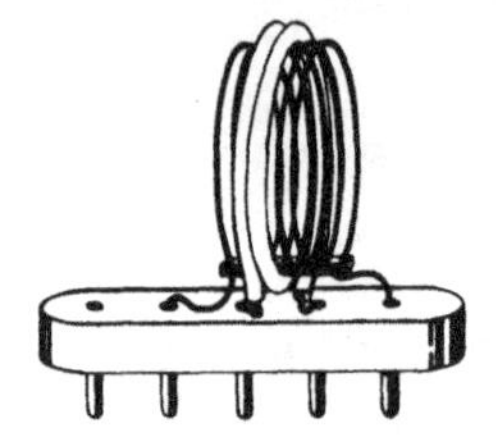

Fig. 1-27. Coil-pin hookup for 15-meter coil.

an *L*-section coupler which will match a 52-ohm coaxial line to a single-wire antenna. The pin-jack arrangement is such that we get a 98′ wire (the feeders are tied together) on 80 meters.

Using the Tuner on 10 Meters

Assume that we are tuning-up on 10 meters first, with the proper coil plugged into the socket. Clip *A*, used for grounding the rotor of the variable capacitor on 80 meters, should be clipped to the rotor. This leaves the rotor above ground, which is the way it should be for all bands except 80 meters. Clips *B* and *C* should be hooked to leads *D* and *E*, which connect to the ends of the coil. The clips are provided so that the tuner

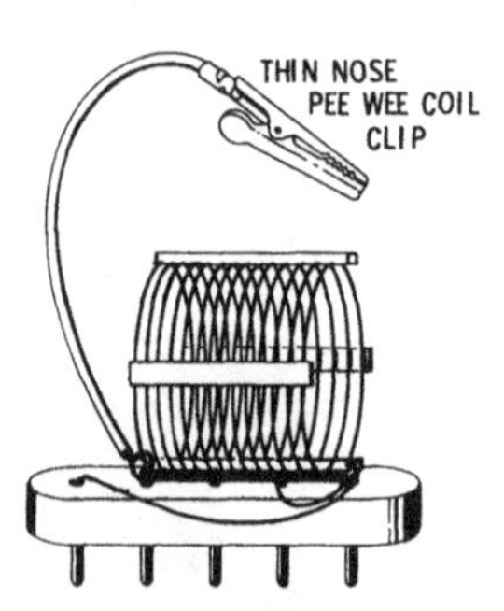

Fig. 1-28. Coil for 80-40 meters.

can be used with feed lines and tune-up procedures which require tapping down on the coil. Now hook the coaxial line from the transmitter to the receptacle on the tuner. Connect the two-wire feed line to terminals *F* and *G*.

As will be discussed in Chapter 3, an SWR meter can be used for adjusting the tuner. However, you can do the job without the meter, by using a trial and error method.

1. Make up a pick-up loop with a #47 lamp as shown in Fig. 1-29.
2. Set the 350-mmf coaxial-line coupling capacitor to approximately ½ capacity.
3. Turn on the transmitter and adjust the output capacitor for low output. Also adjust the plate-tuning capacitor for resonance.

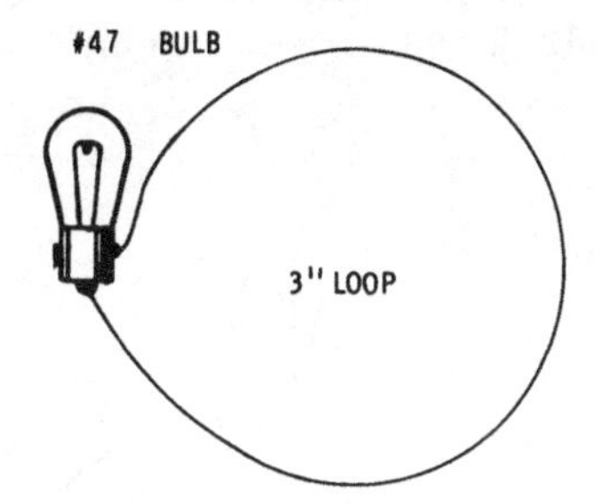

Fig. 1-29. Construction details of the pickup loop.

4. Loosely couple the 3″ pick-up loop to the coil.
5. Adjust the main tuning capacitor for maximum brightness as indicated on the pick-up loop.
6. Adjust the 350-mmf capacitor (at the rear of the chassis) for maximum bulb brightness. To avoid detuning effects, keep the coupling as loose as possible.
7. Readjust the transmitter for resonance and proper load.

Tuning for 15 and 20 Meters

The tuning procedure given above will work equally well on 15 and 20 meters. Simply plug in the proper coil, and follow through as outlined.

Tuning for 40 and 80 Meters

First, plug in the 80-40 meter coil; then ground the rotor of the capacitor by clipping *A* to ground point *11* on the chassis (Fig. 1-24). The best way to tune is with an SWR meter, as described in Chapter 3. However, you can do it with a flashlight bulb or other simple output meter. If your power is under 75 watts or so, hook a number 44 bulb (or small flashlight bulb) in series with the antenna feeder wire which goes to the 68′ flat top. For higher power, tap the bulb over a foot or so of the feeder, as in Fig. 1-30.

Tune up first on 80 meters, with the transmitter set on that band of course. With the tuning capacitor completely closed, try tapping various turns of the coil, starting at the tap end. Probably, within 5 turns or so of the end, the bulb will light dimly. (Readjust the transmitter for resonance.) Now rotate the tuning capacitor for maximum bulb brightness. (Readjust the transmitter for resonance.)

Next try tapping one turn or so either side of the tap portion which first gave an indication of output. Again, readjust the transmitter for resonance. The idea is to get the proper combination of coil and capacitor which gives the most output for a given input as indicated by the meters on the transmitter. Mark the proper tap spot on the coil.

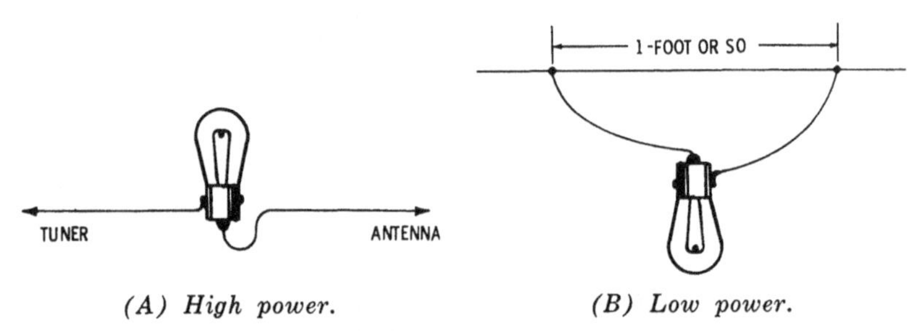

(A) High power. *(B) Low power.*

Fig. 1-30. Resonance indicator.

Tuning for 40 Meters

The tuning procedure on 40 meters is exactly the same as the preceding. However, the coil tap location and capacitor setting will be different. *Important:* Don't forget to remove (or short across) the tuning bulb if it is connected in series with the antenna. Keeping it lighted wastes power! However, if the bulb is simply tapped across a portion of the feeder, leave it alone as the power waste is small.

CHAPTER 2

Basic Vertical Antennas

For many years, in ham radio it was common in over-the-air QSO's to hear some OM comment that, on trying a vertical antenna, he found that not only did it radiate equally well in all directions but also equally poorly in all directions. Every now and then someone would confound the pseudo experts by coming up with a walloping signal from a simple vertical antenna. But nobody really seemed to understand why. Today, we understand why the right kind of vertical antenna can, and does, brew up an effective signal.

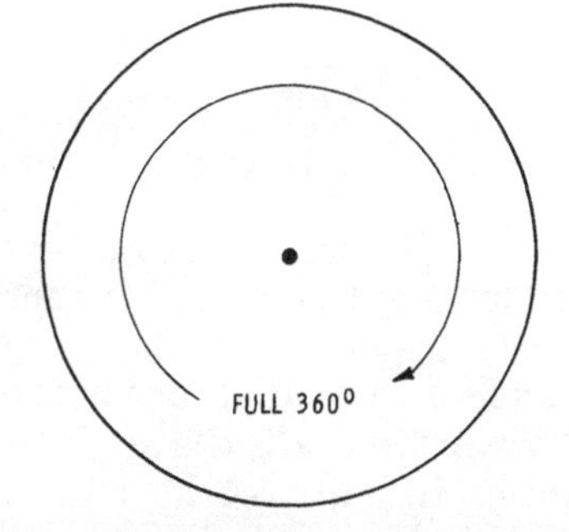

Fig. 2-1. Radiation pattern of a vertical antenna; top view.

One of several things you need to know about vertical antennas is illustrated in Fig. 2-1, which shows a top view of a vertical antenna. Notice that the signal is radiated the same in all directions. This pattern neatly overcomes the problem with horizontal antennas which requires that they be carefully oriented in order to send a signal in the desired direction.

Next, consider the angle of radiation of the antenna. Every antenna has a specific angle of radiation. For example, the radiation pattern of a horizontal-dipole antenna one-half wave above the ground is essentially a doughnut with the power

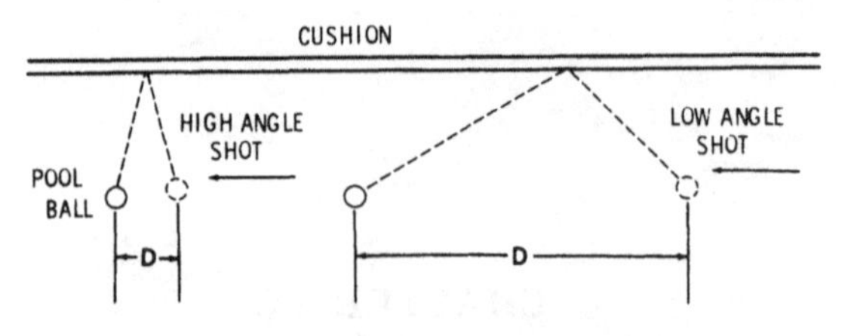

Fig. 2-2. Pool-ball analogy of the angle of radiation.

concentrated at no particular angle, except for the effect of the ground. This is unfortunate, since low-angle radiation is more desirable for long-distance communication.

Fig. 2-2 shows a good analogy—a pool ball bouncing off the cushion on a pool table. A high angle with the cushion brings the ball back close to the point where you shoot it. But if the ball hits the cushion at a low angle with the side of the table, it goes a lot further.

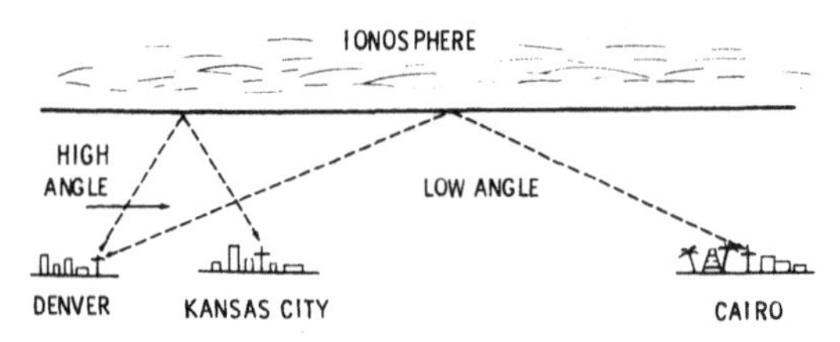

Fig. 2-3. Low-angle radiation for DX.

In radio transmission, exactly the same thing happens. If the transmitted signal is concentrated at a low angle, it strikes the ionosphere at a low angle, leaves it at a low angle, and travels a great distance before coming back to earth (Fig. 2-3). For working DX, the angle of radiation is extremely important. Notice in Fig. 2-4 how the bulk of the radiation is at a low angle to the ground.

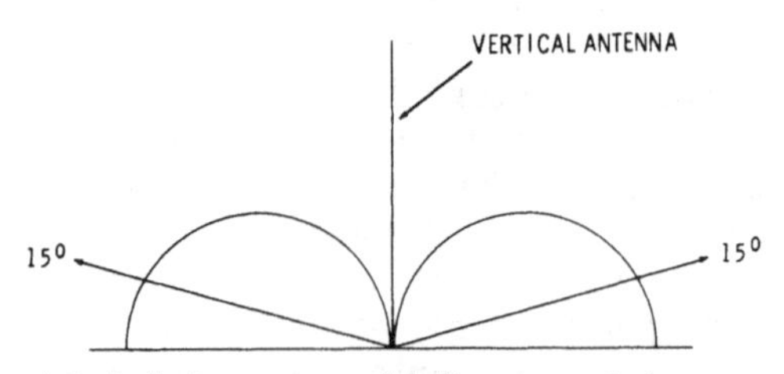

Fig. 2-4. Radiation pattern of half-wave vertical antenna. Note the bulk of radiation at low angles.

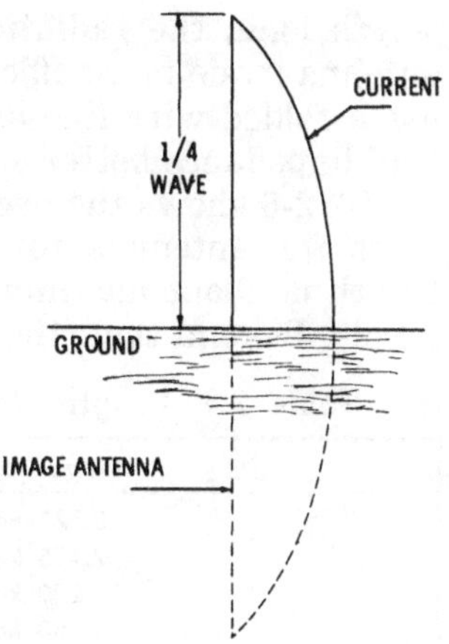

Fig. 2-5. Using the earth as the other half of a vertical antenna.

With vertical antennas, two other things become extremely important. First, vertical antennas do not perform well unless they are over a good ground. Second, except on the three higher-frequency bands, a vertical becomes quite high when used as a half-wave antenna. (For 40 meters a half-wave vertical would stick up in the air 66 feet.)

In practice, the antenna can actually be made a quarter wavelength long, and the ground used to provide a kind of mirror image, which makes up the other half of the antenna (Fig. 2-5). However, this makes the ground efficiency even more important. It is a rare ground arrangement which does not waste even a little power.

Finally, there is another factor you should consider—height above ground. Ideally, the vertical antenna has its low end close to the ground. If you raise it very far, you also raise the angle of radiation. The probable reason that many hams originally had bad luck with verticals is that they assumed that, like horizontal antennas, the higher it was, the better.

Virtually all radiation comes from the bottom of the antenna, which is the high-current point. This frequently means that some of the transmitted signal is soaked up by objects in the area, including water pipes and downspouts, guy wires on power-line poles, and conduit in buildings. Despite the disadvantages, however, a vertical antenna can do a wonderful job. Look at a number of antennas which experience has demonstrated work out well in practical installation.

THE FOLDED-WIRE VERTICAL

In the description of the *Novice*-90 antenna, it was mentioned that, by making the vertical more than a quarter-wavelength long, the radiation resistance could be raised and the antenna made more efficient. This is one method; another is to use a folded-wire line as the radiator. Such an antenna has a load impedance better suited for feeding with a coaxial line.

Fig. 2-6 shows the overall layout. Table 1-1 gives the dimensions for antennas for 7 through 28 megacycles. The 7-mc length is about maximum for the usual vertical supported by a pole. Beyond that, the antenna should be run at an angle, as

Table 2-1. Lengths for the Folded Wire Vertical Antenna

Frequency	Length
3,725 kc	62'-9"
7,125 kc	32'-6"
14,200 kc	16'-6"
21,150 kc	11'
28,700 kc	8'-6"

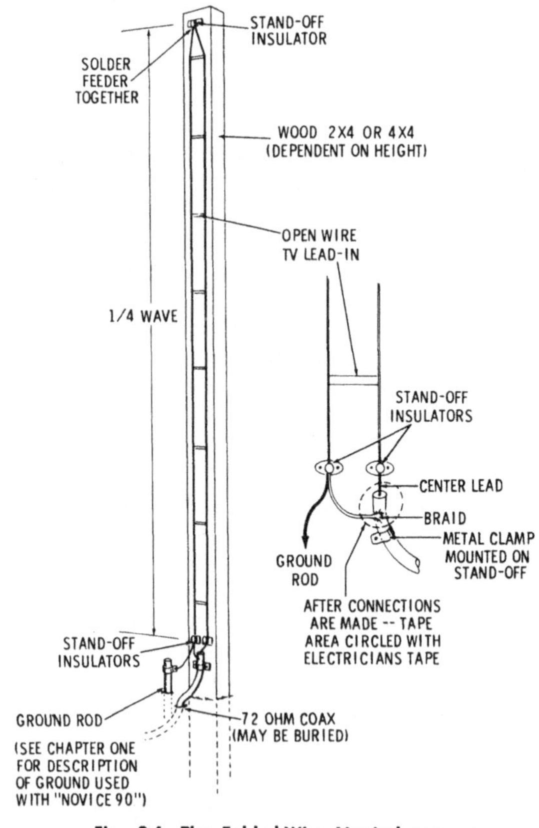

Fig. 2-6. The Folded-Wire Vertical antenna.

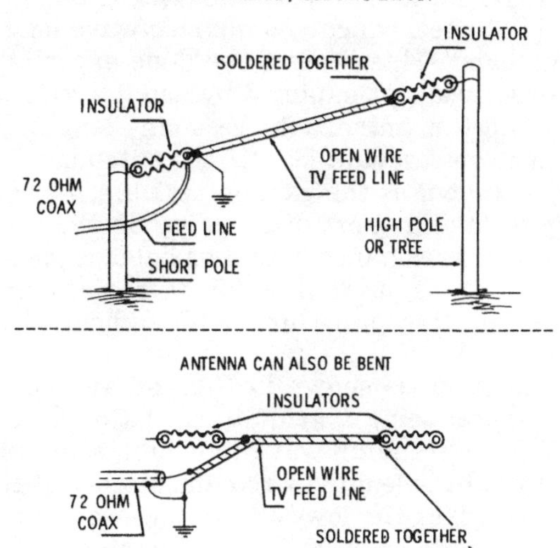

Fig. 2-7. Dimensions for antennas of from 7 to 28 mc.

shown in Fig. 2-7. Note that, in every case the antenna is fed with a 72-ohm coaxial line.

COAXIAL LINES

The younger generation of hams is apt to regard the coaxial line as the living end. In fact, many of them do not have any clear idea that you can feed an antenna with any other type of line. Actually, coax is anything but an unmixed blessing. Unless the standing-wave ratio is kept low, losses can be very high. This subject is covered in more detail in the chapter on using antenna tune-up equipment, which explains how to go about keeping the standing-wave ratio low. Losses on an open-wire line are much lower; such a line has a lot to recommend it for many amateur applications.

For verticals, though, it usually works out best to use a coaxial line. So, the thing to do is to buy the best line you can find. Always avoid surplus and bargain lines. They invariably have a high loss which gets steadily worse as the line ages.

THE HOT-POLE RADIATOR

If, as was suggested earlier, you are using a TV-type adjustable mast to support your horizontal antenna, you can, by getting the length right, use the mast as a 40-meter vertical for practically no additional cost. The trick is to get the height approximately 33 feet, which is a quarter-wave long in the 40-meter (7-mc) band. This needs to be done experimentally, by the method described in Chapter 3 by using a grid-dip oscillator for adjusting an antenna to resonant length.

Once you achieve resonant length, the antenna can be driven by a technique known as the gamma match. This consists of a wire tapping located a short distance up on the radiator, and then tuning out the reactance with a variable capacitor. Differing ground conditions, as well as in the radiator itself, make exact detailing of this procedure a bit difficult, but in brief here is the method.

Set up the antenna as shown in Fig. 2-8 and adjust the antenna to resonance, using a grid-dip oscillator as described in Chapter 3. With a standing-wave indicator some place in the feed line, work out the length of rod and setting of the variable capacitor which gives the lowest standing-wave ratio (SWR) using an SWR meter as described in Chapter 3. The two adjustments are interrelated, so experimenting will be necessary.

Is there interference with the antenna supported by the mast? Ordinarily this will be a small factor, because horizontal and vertical polarizations do not mix very much. Of course, it is important that the guy wire be separated into sections by insulators.

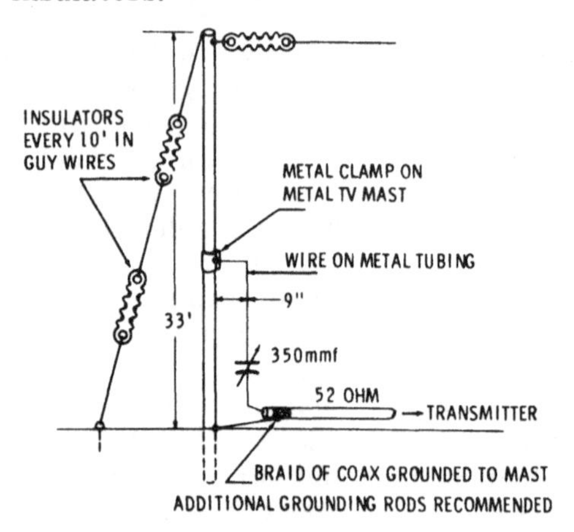

Fig. 2-8. The Hot-Pole Radiator for 40 meters.

80-METER LOW-ANGLE RADIATOR

The vertical antennas described so far are best suited for 7 mc or higher because the overall length becomes too great for a 3.5-mc antenna. However, by some loading technique, it is possible to build an effective 80-meter vertical without having a lot of height.

The problem of building such an antenna has been greatly simplified by the availability of the 23-foot self-supporting all-band antennas currently manufactured by at least three different companies. The reason for the 23-foot length is that anything longer results in an antenna which does not give good low-angle radiation on 28 mc. However, the short length penalizes operation on 80 meters; like most all-band antennas, there have to be some compromises.

However, by sacrificing the all-band feature, it is possible to build an 80-meter antenna with a minimum of trouble. At very long distances, this antenna will perform better than any horizontal 80-meter antenna under 50 feet in the air; and probably better than one which is considerably higher than that.

Commercial antennas are actually a vertical radiator of tubing, plus a loading coil. The latter makes up for the lack of length needed to provide a quarter wavelength on 80 meters. Of course, a coil radiates relatively little—and every bit of vertical radiating surface that can be added will help make the antenna more efficient. Therefore, with this in mind, mount the antenna as high as is practical, and then piece it out with some aluminum wire, thus adding to the effective length.

Fig. 2-9 shows the overall layout of the antenna system. As indicated, it can be placed alongside a house, or on a pole. In either case, make the overall height as great as you can; anything up to 60 feet will work fine.

If you are mounting the antenna either against a house or on the side of a pole, the first step is to provide a firm base support. This can usually be done by fitting the aluminum tubing with a couple of small straps that are secured to the heaviest stand-off insulators you can buy. Strong insulators are a must; the forces which develop at the base in a wind are considerable. It may be that with some single modification you can take advantage of the insulators that come with the antenna; the important thing is that the installation be strong.

From the end of the antenna tubing, run some heavy aluminum wire (like TV antenna ground wire) down to an insulator mounted on a 4 × 4 (one end of the 4 × 4 either buried in the ground or in concrete). The wire should terminate at a stand-off insulator. The coil provided with the antenna now becomes a loading coil at the base of the lengthened antenna (Fig. 2-9), and is mounted between the insulator on the antenna lead wire and another insulator. This lower end of the coil is fed via a coaxial line, as shown in Fig. 2-10. Note the detail of how to get a good insulation job on the end of the coax. Of course, the job can also be done very neatly with a coax connector of the type illustrated in Fig. 2-11. Note that the braid side of the coax goes to ground. This ground is an extremely important element; so important, in fact, that you can not possibly get one as good as you would like.

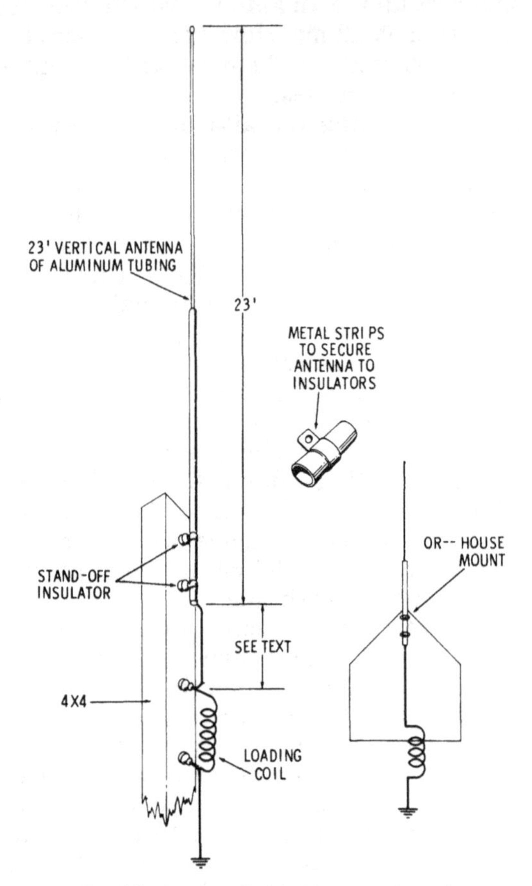

Fig. 2-9. Layout of the 80-Meter Vertical.

Getting a Good Ground

The simplest way to provide a fairly good ground is to use the chemical-ground approach, as illustrated in Chapter 1. Since, for an 80-meter vertical, the ground is far more critical than for the *Novice* 90, it is a good idea to drive at least three ground rods and hook them all together. An even better method

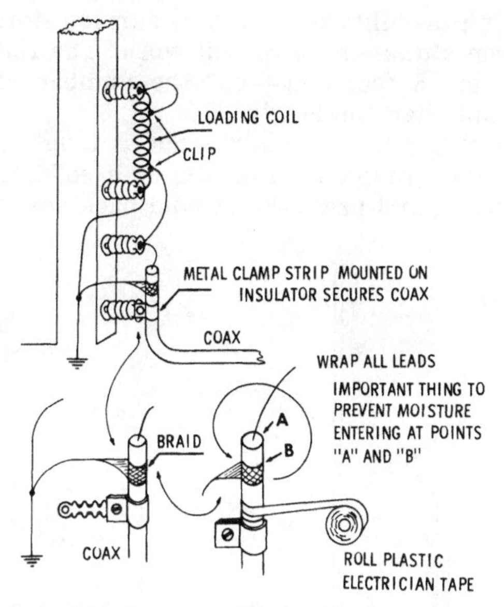

Fig. 2-10. Connecting the coaxial line to the coil.

is to supplement the ground rods with a fan of radials, buried an inch or so under the surface of the ground. This is the approach used by broadcasting stations, who often will put out hundreds of radials. Of course, this is impractical for the ham, but put out as many as you want to and can (Fig. 2-12). For low-priced wire, try an electrical motor repair shop. You may be able to find some motor windings with long lengths of usable

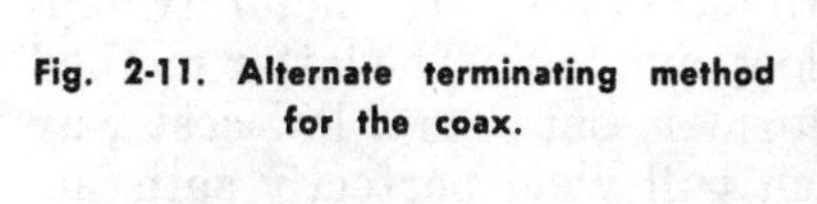

Fig. 2-11. Alternate terminating method for the coax.

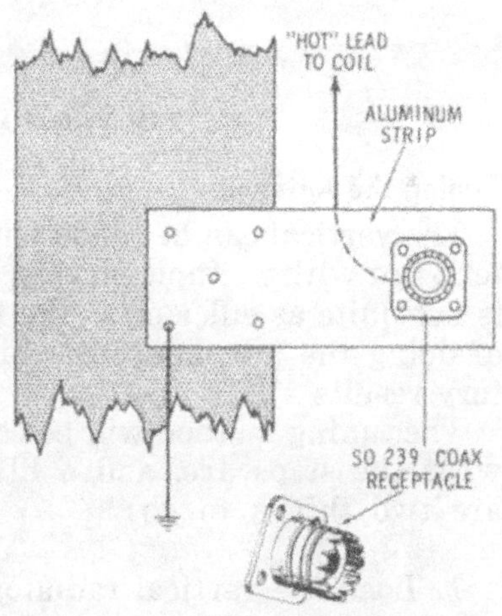

wire. Another possibility is a military surplus store. Most any copper or even aluminum wire will work. The radials can be anything up to 66 feet long—but the number of radials is more important than the length.

Actually, putting down radials is not as tough as it sounds. You can slit the ground with an ax, or a sodding tool, poke the wire in place, and press the ground back down.

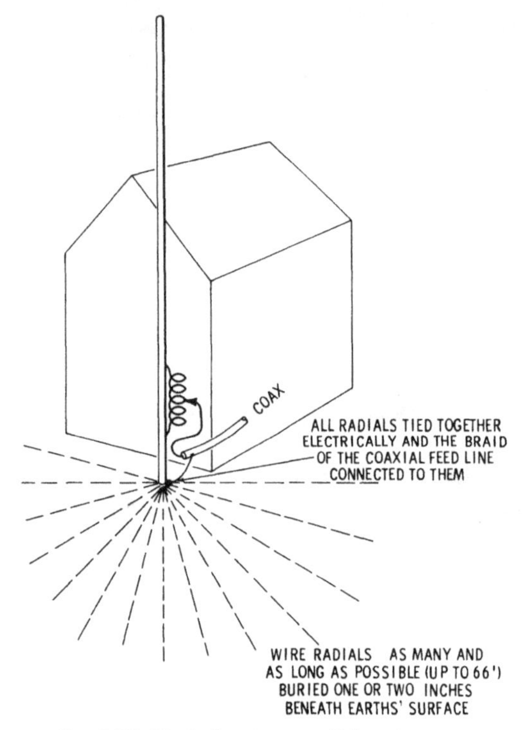

Fig. 2-12. Vertical antenna radial system.

Tuning the Antenna

The vertical can be tuned with either an SWR meter (Chapter 3) or with a simple pick-up loop and bulb. The latter method is not quite as efficient as the former, but it is a low-cost way of doing the job; and one which will yield perfectly satisfactory results.

The tuning method will be easier to follow if you understand what the steps are, and a little of the why. Actually, there are two things to do:

1. Load the vertical radiator with sufficient inductance (by

means of a coil) so that it will resonate at the desired frequency.
2. Match the 52-ohm feed line to the base of the antenna.

In mobile antenna installations this is usually done by simply feeding the bottom end of the loading coil with the coaxial line and then adjusting the coil length for resonance. Fig. 2-13 shows the hook-up. Sometimes the same arrangement is used for commercial vertical antennas.

The method has the advantage of simplicity, but it does not always allow a good impedance match; hence SWR is high. This is of little consequence with an auto antenna since the length of coax is very short. But for a permanent installation, a better match is highly desirable, both to reduce losses in the coax (high SWR wastes power) and also to make the antenna less critical as to frequency.

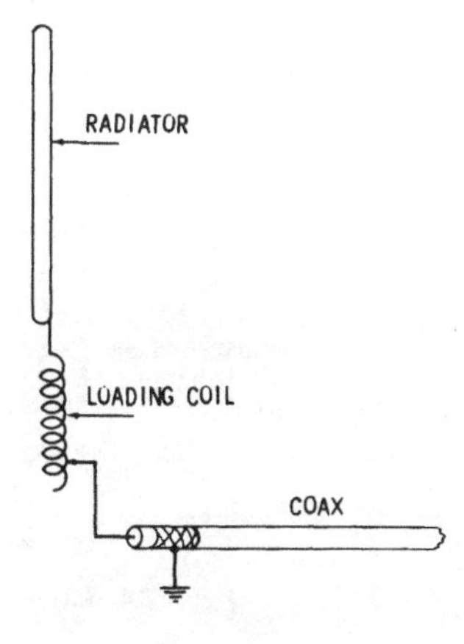

Fig. 2-13. Method of tuning the antenna.

Impedance Matching With A Base Coil

Fig. 2-14 shows the circuit; note that the coil is connected between the antenna and ground, and the length of the coil is adjustable. The feeder line is tapped onto the coil at a point which gives a match for 52 ohms, the coil, in effect, serving as a tapped matching transformer, as well as a loading coil. Note the steps:

1. Place coil clip *A* on about the eighth turn from the grounded end.
2. Fire up your transmitter on 80 meters and adjust for input of approximately 50 watts. Be certain that the plate-tuning capacitor is adjusted for resonance.

3. Couple the pick-up loop to the bottom end of the coil.
4. Adjust coil clip *B* for maximum brightness, as indicated by the pick-up loop.
5. Adjust clip *A* to further peak brightness. Always use the maximum number of turns which gives the same brightness. For example, if tapping at turn No. 10 seems to give the same brightness as with clip on turn No. 8, use turn No. 8.
6. As a final step, readjust tap *B* for maximum indication on the pick-up bulb.

As you may have figured out, the *B* adjustment is for resonance, and the *A* adjustment is for impedance matching to the coax feed line.

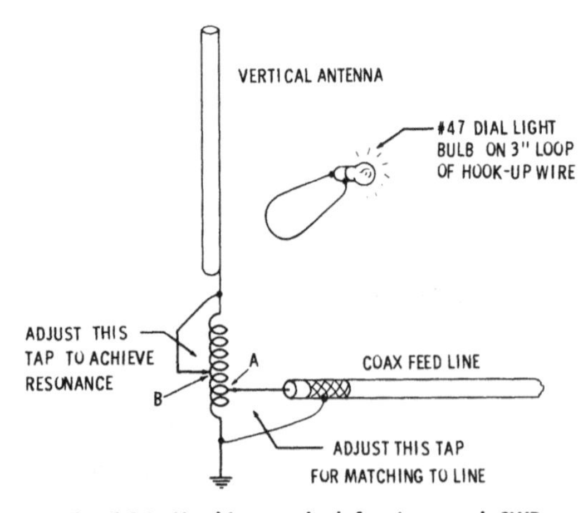

Fig. 2-14. Matching method for improved SWR.

THE "33-OUT 23-UP" FOR 40 METERS

At certain sunspot cycles, every band below 7 megacycles goes pretty sour, and hams who have been used to the easy DX on 21 and 28 megacycles all at once have to back down to 7 megacycles. Getting a good low-angle radiator at that frequency can be quite a problem.

Even the big trap beams usually are not as good as you'd like on 40 meters because it is so difficult to get them high enough: for low-angle radiation a tower one-half wave high (66 feet) is not a bit too much—yet it is hard to come by. For the average ham with a limited pocketbook, a vertical antenna is usually a more practical approach.

As mentioned earlier, a vertical usually needs a good ground system. However, some years ago, West-Coast hams popularized a type of antenna called the "33-out 33-up" which does not need a ground. Actually, it is the close relative of a ground plane to be described later; but it is much simpler than a ground plane, low in cost, and easy to put up. Further, by using one of the commercial 23-foot verticals mentioned earlier, it can be made self supporting and quite neat. The author calls this new version the "33-out 23-up."

In many ways the antenna is like a dipole, with one portion vertical and the other horizontal. In the old days, on 160 meters, such an antenna was called an *antenna-counterpoise.* Actually, the single-wire horizontal portion does not provide a counterpoise in the strict technical sense, but it does improve the overall efficiency of the system.

This horizontal leg is ordinarily mounted at approximately 7 feet off the ground—high enough not to knock your hat off, but low enough to leave most of the transmitting up to the vertical portion. Run the wire out into the backyard like a high clothes line, or mount it on insulators around the base of the house.

You can lay out such an antenna in two different ways, as illustrated in Fig. 2-15. The method shown in Fig. 2-15A is more compact; however, the other (Fig. 2-15B), using a full 33 feet of vertical radiating surface, is more efficient.

Tuning the Antenna

If a full 33 feet is used for both legs, no tune-up is required; the antenna simply acts like a coaxial-fed dipole. The 33-23 version is also simple to tune up; all you have to do is to tune the entire system to resonance, and this can be done in one of two ways.

1. With a pick-up loop made of a loop of wire and a dial lamp fastened to a stick (Fig. 2-14), couple the coax between the base loading coil and the 33-foot counterpoise leg. With the transmitter set to the desired frequency, simply adjust the length of the loading coil for maximum output as indicated on the pickup loop. (Turn the transmitter off each time there is an adjustment to make.)
2. If an SWR meter is available, adjust the coil for minimum SWR (Chapter 3). By using an SWR meter, a somewhat more accurate method is obtained than by just using a pickup bulb.

THE SNEAKY GROUND PLANE

The idea of obtaining low-angle radiation from a vertical antenna high enough in the air to be in the clear of obstacles has held a certain fascination for hams for years. Fortunately, there is a way to do it—with an antenna called the *ground plane*. In effect, wire radials are used to establish an artificial ground so that the radiating portion can be high; for example,

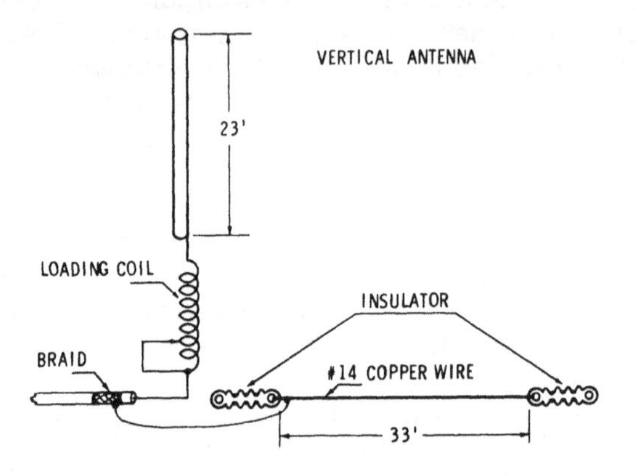

(A) 40-meter DX antenna.

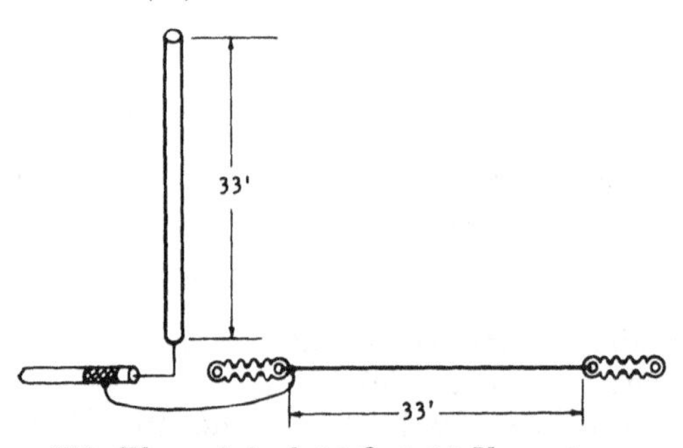

(B) The original 33-Out 33-Up antenna.

Fig. 2-15. The 33-Out 23-Up.

on top of a house. Here is one version of such an antenna which the author has used successfully; it is called "the sneaky ground plane."

First, notice in the overall layout of a ground plane (Fig. 2-16) how the ground lead of the coax goes to the radials. The base impedance of a ground plane is in the neighborhood of

30 ohms, which is a bit low for a good match with a 50-ohm line. Various ways of achieving better match have been worked out; some of which will be covered later. Fortunately, a better match can be obtained by a technique which, in addition, usually makes erection simpler. That is, by allowing the radials to droop. This, in effect, raises the radiation resistance. Now, the sneaky part.

Most of the antenna radiation comes from the vertical portion. The radials, therefore, can be hidden under the roof if

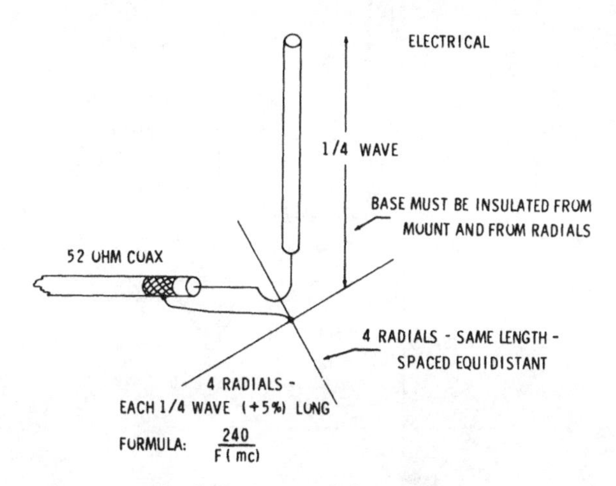

Fig. 2-16. A ground-plane antenna.

desired. In one 28-mc installation, the author did exactly that. The only thing visible was an 8-foot whip on top of the house. The radials were in the attic, as shown in Fig. 2-17. Such an installation has possibilities in many homes.

Ideally, there are four radials. However, fairly good results can be obtained with three, and this works out particularly well for the end of a house, as shown in Fig. 2-18.

The radials should be calculated using the formula 240/f(mc). Table 2-2 will give you the lengths for roughly the center of each of five bands for the ground plane.

THE GUTTER-SNIPE GROUND PLANE

Actually, almost any fairly large piece of metal will serve as a ground plane of sorts. If you have a large gutter, you can tie onto it for the ground plane and use a vertical radiator. Such an antenna is almost certain to radiate better in some directions than in others. But even so, it may work surprisingly

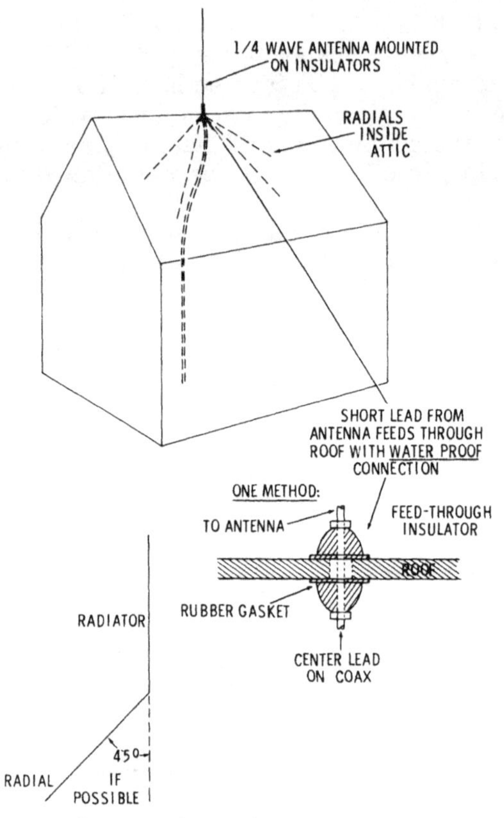

Fig. 2-17. The Sneaky Ground Plane.

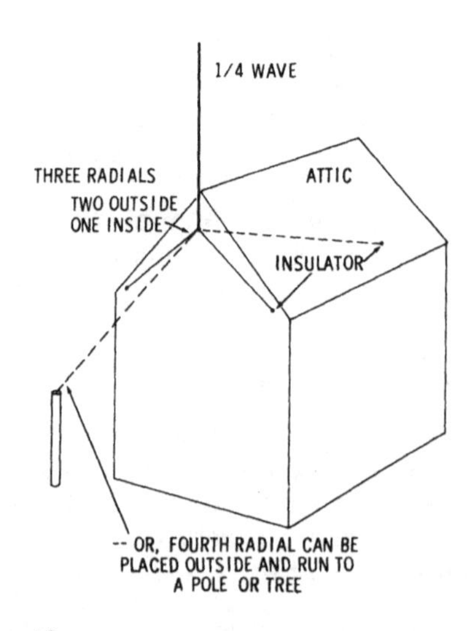

Fig. 2-18. A three-radial, ground-plane antenna.

Table 2-2. Ground-Plane Antenna Radial Dimensions

Band	Length of Radial
80 meters	66′- 8″
40 meters	33′- 5″
20 meters	16′-10″
15 meters	11′- 4″
10 meters	8′- 8″

well. It is so easy to build that it is well worth the effort if you have room for nothing else (Fig. 2-19). Getting a match is difficult because every gutter will perform differently as a ground plane. However, if your feed-line is short, the standing-wave ratio should not be too bad just as shown. And, by using an SWR meter and a radiator .32 wavelength long, you can achieve a better match, as will be discussed in Chapter 3.

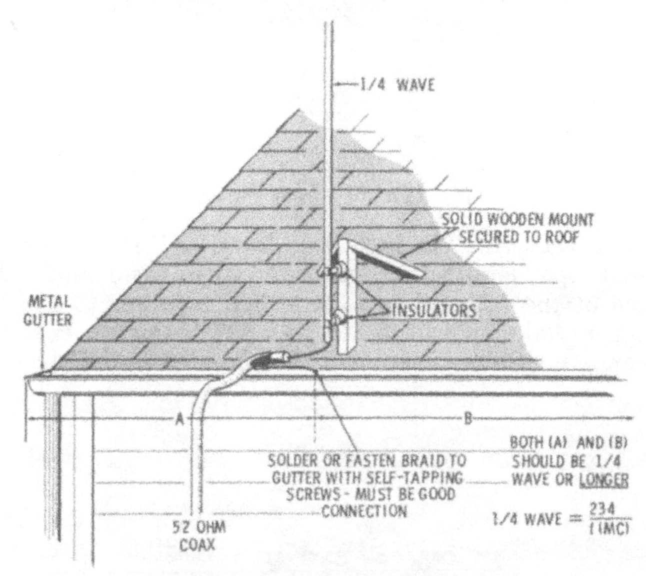

Fig. 2-19. The Gutter-Snipe Ground Plane.

THREE-BAND VERTICAL

The ground plane can be modified to provide a multi-band, low-angle antenna which does not require tuning in changing bands. One practical way to do the job is shown in Fig. 2-20.

Note that the radiator is a trap-type vertical, which is best bought ready made. The four radials can be made by cutting off the proper lengths of wire from the four-wire cable used to

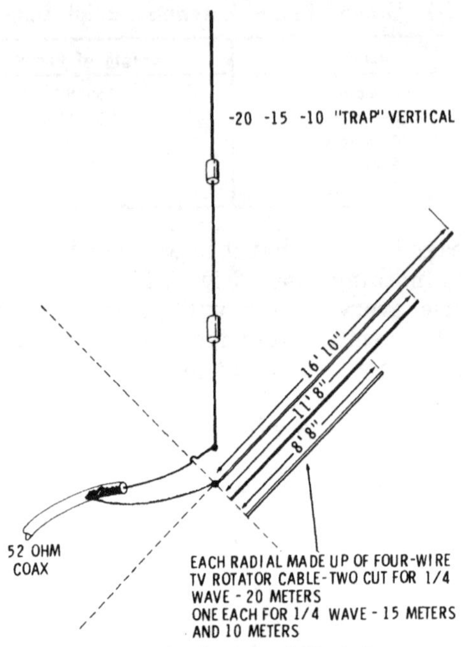

Fig. 2-20. The 3-Band Vertical.

supply TV rotators. The idea is to have each radial include a wire the proper length for each of the three basic bands. Actually, two of the four wires can be tied together to make the 20-meter radial, and the other two cut to length separately for 15 and 10 meters. These can be concealed in the attic or, if

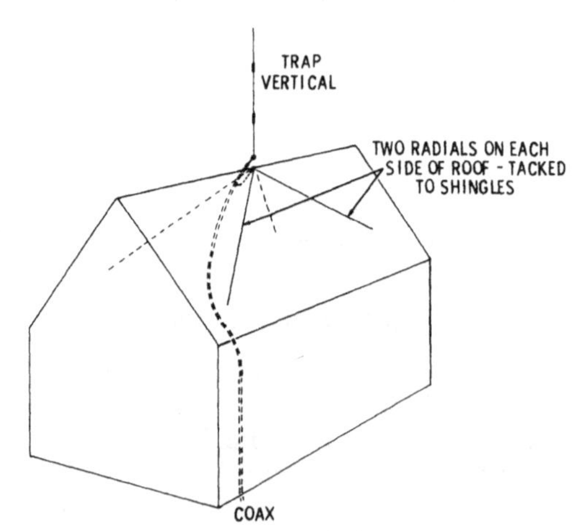

Fig. 2-21. The 3-Band Vertical attached to the roof of a house.

easier, simply tacked on the roof of the house, as shown in Fig. 2-21.

A HOME-MADE FIVE-BAND VERTICAL

For the ham with very little space for an antenna (and who is also interested in low-angle radiation for extreme DX) a simple five-band vertical has some real advantages. Commercial antennas of this type are available from several companies, and cost less than twenty dollars. However, you can build your own; and while it won't be as neat and trim as a manufactured antenna, if carefully made, it will yield good results.

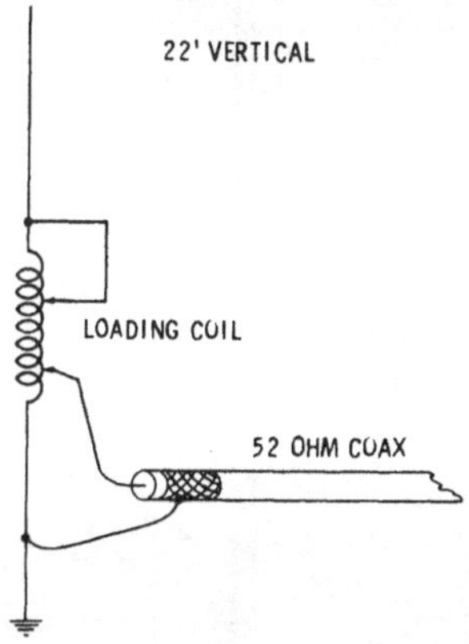

Fig. 2-22. A homemade, 5-band vertical antenna.

Note in Fig. 2-23 that the radiating portion is 22 feet. As was mentioned earlier, the reason for this is that for low-angle operation on 10 meters, this length is approximately ¾ wave—the maximum length which will yield low angle radiation. (Commercial antennas, which use smaller tubing, are ordinarily 23 feet long.) To achieve resonance on the lower frequencies, the radiator is loaded with a coil. The coil also allows the feed line to be tapped in at the point which provides a good impedance match with the feed line.

Fig. 2-23 shows the mechanical layout for the antenna, also the various parts involved. Note that the radiator itself is made from aluminum TV masts. Two 10-foot masts, plus one 5 footer, can be combined to achieve the desired length (cut off the overage with a hacksaw). After the mast sections are forced together (they are tapered to make this possible) secure them with self-tapping screws. The idea is to get a 22-foot pole solid enough to be self supporting.

The most difficult part of the job is to secure a strong base support; in a wind the stresses become quite high. Mount the vertical pole with strips of aluminum, which clamp down tight on the mast tube when the clamps, in turn, are secured to the behive insulators. The latter should be the short, fat type; the heaviest you can find (Fig. 2-9). The coil can be one

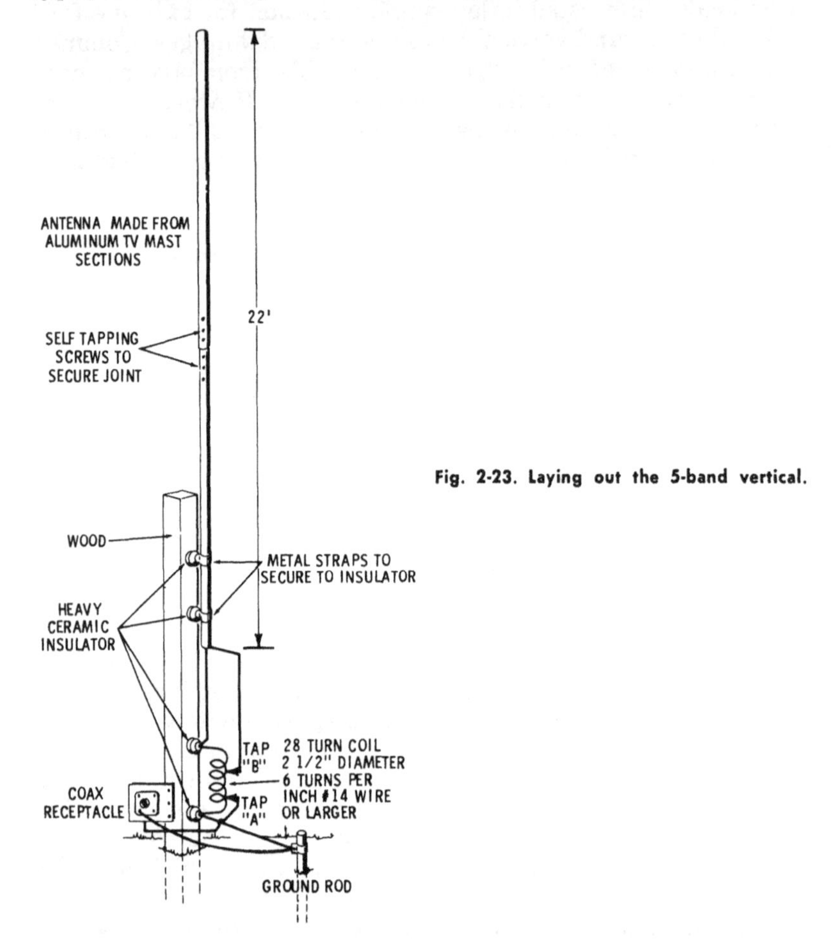

Fig. 2-23. Laying out the 5-band vertical.

you make up yourself following the method outlined in Chapter 1, or can be commercial coil stock. The coil should be mounted securely; ideally, cemented down with china cement to a strip of clear plastic to serve as a base support.

Make the coil of bare copper wire (#14) so that the turns can be easily tapped. Choice of tapping clips is important. You want one which has flat ends so that it will slip between turns and tap one turn without shorting out others. Mueller clip number 88 is ideal.

As with all grounded antennas, the quality of the ground is most important, especially on 80 and 40 meters. Use one of the techniques discussed earlier in this chapter for achieving a low-resistance ground.

Tuning the Antenna

The best way to tune an antenna of this type is with an SWR meter; the method is covered in detail in Chapter 3. However, a satisfactory tune-up can be achieved with a simple pick-up loop, following the same basic procedure as was outlined for the 80-meter vertical.

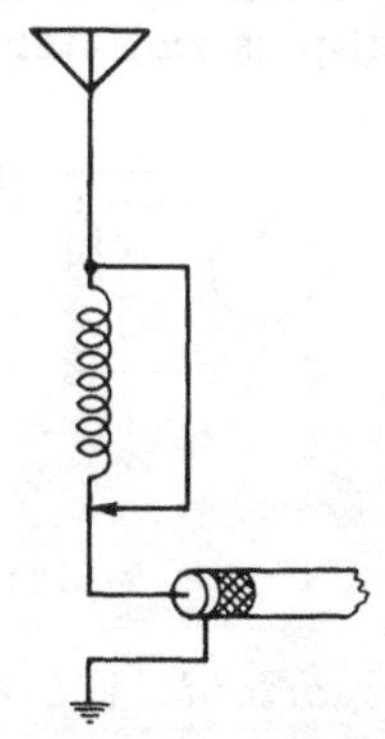

Fig. 2-24. Shorting out the coil.

On 10 meters, the antenna length is so close to a resonant length that ordinarily no tuning is required. The coaxial braid line is simply connected to ground, and the coil shorted out (Fig. 2-24). However, if the antenna does not seem to take power satisfactorily, try adding one or two more turns to the coil.

On the other bands, the procedure is a bit more complicated, but simple enough if you follow through in an orderly, step-by-step way.

1. Set tap *A* from the coax as shown in the following listing: 80 meters—7 turns from grounded end; 40 meters—5 turns; 20 meters—3 turns; and 15 meters—2 turns.
2. Get someone to help you by staying with the transmitter; both to turn it on and off, and also to keep it tuned to resonance.
3. Apply coil tap *B* to the coil. For 80 meters approximately 24 turns; 40 meter—12 turns; 20 meters—6 turns; and

for 15 meters—3 turns (these turns are between the tap and grounded end).

4. Hook the feeder to the transmitter. Turn the transmitter on and adjust the output tuning capacitor to a relatively low output (one-half on a small rig; not over 50 watts or so on a big one).
5. With the power on, couple the pick-up loop to the loading coil. Use the loosest coupling which will give an indication on the bulb.
6. Have your helper at the transmitter turn the transmitter off. Move clip *B* up one turn. Turn the transmitter on and see if output has come up. If not, try going down one turn. The object is to adjust clip *B* for maximum output as indicated by the bulb. This will be achieved when the coil-plus-radiator are on resonance.

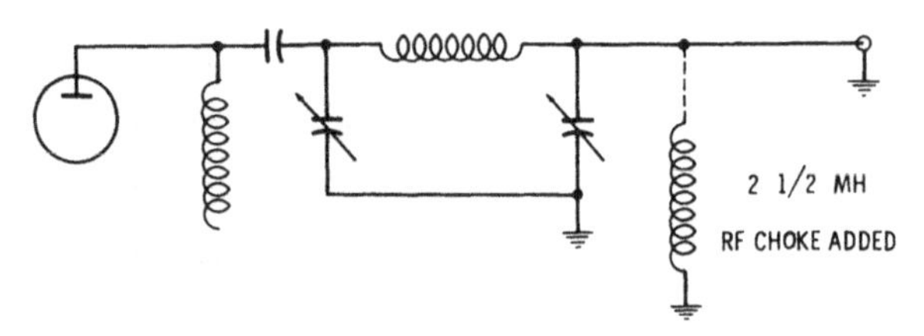

Fig. 2-25. RF choke across the output.

Important: the tuning procedure will be speeded up a great deal if you work with the circuit live and quickly run the clip up and down the coil to achieve the desired maximum indication on the bulb. However, there is a certain inherent risk, since, if the blocking capacitor in the transmitter breaks down, it could place plate voltage on the output. One way to eliminate this problem is to clip an RF choke across the cable. In case of a breakdown, the RF choke will provide a path to ground, as shown in Fig. 2-25.

7. The next setting is for tap *A*, which is not as critical. Adjust for maximum output as before. Always use the maximum number of turns, if there is a choice. For example, if output seems to be the same with four turns and with six turns, use six turns.
8. A final touch-up of tap *B* may increase output slightly.

Once the proper points have been found on the coil, you may want to mark them in some fashion so that you can put the clips back in the proper position for changing bands.

WHICH ANTENNA: VERTICAL OR HORIZONTAL?

At this point you may be pretty confused as to whether you want to use a vertical or a horizontal antenna. Assuming you want to work both nearby and distant stations on 80 and 40 meters, the best answer is *both.* For close-up work on 80 meters for example, a high-angle horizontal antenna (and, on 80 most horizontal antennas are high angle because of the difficulty of getting enough height) will knock the socks off any vertical. This will usually hold true up to 1,000 miles or so. Beyond that, for an indeterminate distance, the two antennas may be about on a par. Beyond 1,500 miles, generally, the low-angle vertical antenna will be more effective.

The most effective distances will be somewhat different on 40 meters, but the same principles apply: use a horizontal antenna for short haul, and a vertical for long haul DX. For 20 meters, where low-angle radiation becomes extremely important, unless you have the money for a fairly high mast (at least 45 feet), a vertical is a good choice. On 21 and 28 megacycles, either vertical or horizontal antennas properly installed give excellent results.

Important: there is a common (and wrong) idea that if the transmitting antenna is vertical, the receiving antenna should also be vertically polarized. This is true only on local groundwave transmission. Once a signal is bounced off the ionosphere, as happens in all sky wave transmissions, the signal becomes *both* vertically and horizontally polarized. As such, it will be received equally well on either a vertical or a horizontal antenna.

CHAPTER 3

Using Antenna Test Equipment

The problem of antenna tuning has been greatly simplified in recent years by the availability of various types of antenna test equipment. However, hams with real know-how were satisfactorily tuning antennas years before the current crop of test gear had even been invented. The techniques developed then are highly useful today; both as a preliminary method before using commercial test units, and as a practical way for the ham with a thin pocketbook to get on the air at low cost. So, before considering the more sophisticated test techniques, first of all consider some simple and practical techniques from past years.

TUNING UP WITH BULBS

The use of a flashlight bulb or a dial-light bulb in a loop has already been discussed in previous chapters. For certain specialized applications, such a technique is most useful. There are also some other ways to use bulbs, some of which were discussed briefly in Chapter 1.

Measuring Current

Bulbs operate on current, and hence provide an easy means for measuring current. In the tune-up of both the *Novice*-90 and the five-band tunable antenna, a bulb in series with the antenna makes it easy to achieve resonance on 80 and 40 meters. On 20, 15, and 10 meters a bulb in a pick-up loop achieves the same thing on the five-band antenna by providing a method of coupling to the tuning coil, and thus measuring induced current.

For powers up to 150 watts or so, flashlight bulbs or dial lamps will usually handle the current. If such is not the case, however, more bulbs can be paralleled. Also, as shown in

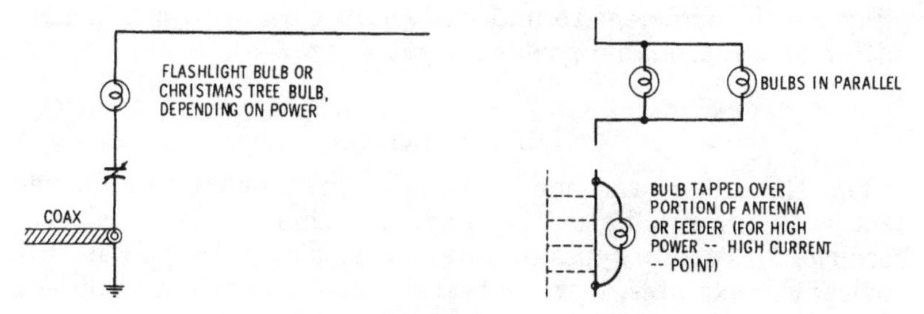

Fig. 3-1. Using small bulbs to measure power output.

Fig 3-1, a bulb can be tapped over a portion of an antenna or feed line. In addition, you can use a Christmas-tree light bulb, in the case of a transmitter with high output.

A filament-type bulb will give the most indication at a current loop, and most antennas are fed at high current (low voltage) points. A grounded (Marconi) antenna invariably is fed at such a point. However, sometimes an antenna is fed at a high-voltage point, and because of the length of feeder in use, the feed point at the end of feeder also represents a high-voltage, low-current position.

Measuring Voltage

In Fig. 3-2 there are actually two ways to get a voltage indication. One is with a pick-up loop; the other, as shown, is with a neon bulb. The latter has the advantage of drawing so little current that it can be left in the circuit at all times, and is less apt to upset the tuning adjustment.

Notice that the leads from the neon bulb are simply connected to the hot side of the variable capacitor. A small metal plate, made from a scrap of aluminum is grounded and simply placed near, or touching, the glass bulb. This capacity cou-

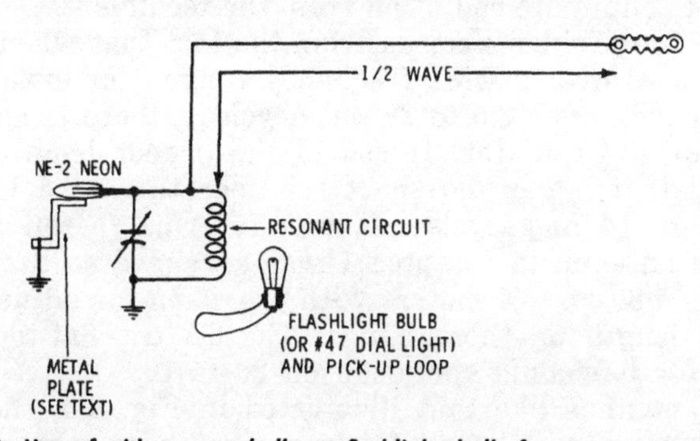

Fig. 3-2. Use of either neon bulb or flashlight bulb for power measurement.

pling will be sufficient to make the bulb glow with any transmitter of a reasonable power output (Fig. 3-3).

RF METERS

The RF meter was once a popular instrument, but in recent years it has not seen too much use. This is indeed a shame because, for certain types of antennas it is an extremely useful device. For example, it is the best device for achieving tune-up of the 5-band antenna on 20-15-10.

This antenna, *on these 3 bands,* is a Zepp; so-called because it was originally worked out for dirigible antennas. Such an-

Fig. 3-3. Construction of neon-bulb method of power measurement.

tennas have seen less use than they deserve because of the difficulty of achieving a good feeder balance, which is desirable in order to eliminate radiation from the feedline.

One of the problems comes from the fact that when the antenna is used over a wide frequency range (for example, for all ham bands from 3.5 to 28 megacycles), there is no correct length for the radiator. If cut to the proper length for 3.9 megacycles, it is far too short for operating as a harmonic antenna on 14 megacycles. It was for this reason that the five-band antenna in Chapter 1 was arranged so that it is a Marconi on 80 and 40 meters with provision for adjusting the resonant length on those bands. The 68-foot flat-top length is ideal for harmonic operation on 20-15-10.

With antennas like that illustrated in Fig. 3-4, the length of the antenna must be cut fairly accurately to achieve a good

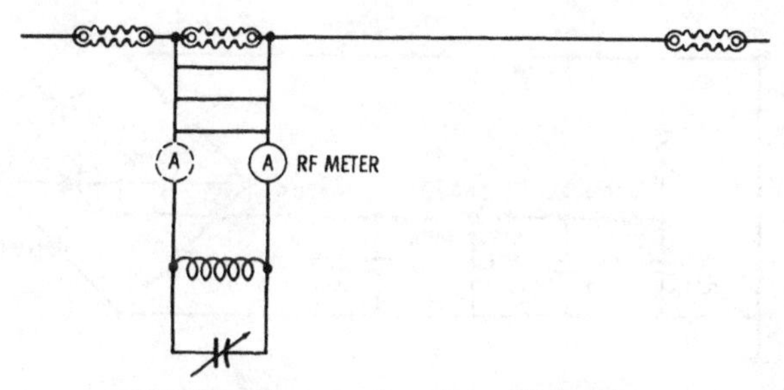

Fig. 3-4. RF meter in series with lead-in.

balance of feeder current. The best way to do this is to utilize an RF meter, and measure the current in *each* feeder alternately. The antenna should then be shortened or lengthened in order to achieve the best balance between the feeders.

The meter can be any RF ammeter in the 0-1 to 0-5 range. The 0-1 range, of course, is for low power, and the 0-5 for high power. Such meters can be bought new, or you may be able to find one in a military surplus store.

Caution: regardless of the range of your meter, in the initial tuneup, jumper the meter with heavy wire and try it out. If there is no indication, reduce the size of wire jumper. You may find that the meter can be used without a jumper; but start out with one to avoid burning out the meter with too much RF (Fig. 3-5).

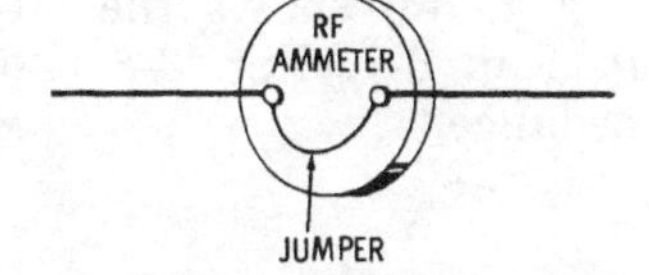

Fig. 3-5. Jumper the meter with heavy wire for primary tests.

The RF meter can also be used to excellent advantage in tuning up the Novice-90 and similar single-wire antennas. The meter, of course, is simply used as a substitute for a bulb in series with the antenna; it has the advantage of giving a precise reading as contrasted with that of estimating the output on the basis of the brilliance of a bulb.

RESISTANCE-BRIDGE SWR METERS

The resistance-bridge type of RF and SWR meter has been supplanted in many cases by the easier-to-use reflected-power SWR meter. However, the bridge will do some things which can not be accomplished with the reflected-power meter. As

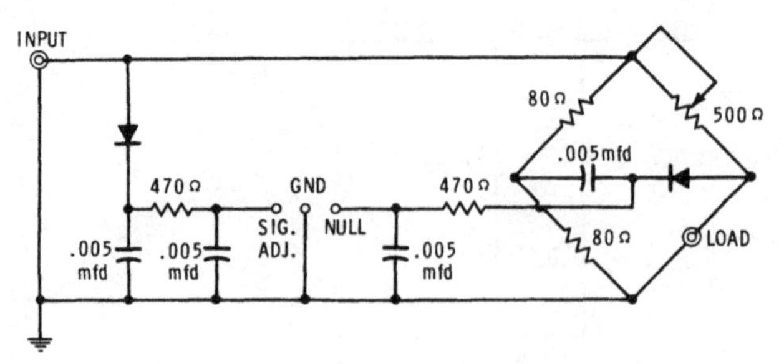

Fig. 3-6. Circuit of a resistance-bridge SWR meter.

an example, the bridge provides a practical way for measuring the impedance of an antenna. Such information can be highly useful in working out a feeder system, for example.

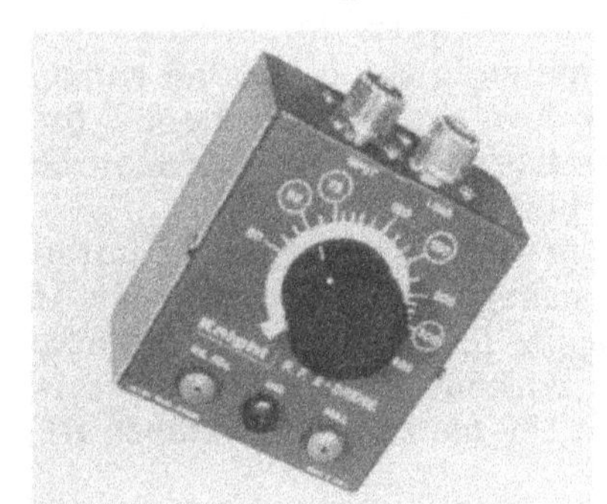

Fig. 3-7. Resistance-bridge SWR meter.

Fig. 3-6 shows the circuit of a typical bridge, illustrated in Fig. 3-7. Fig. 3-8 shows the hook-up for measuring impedance.

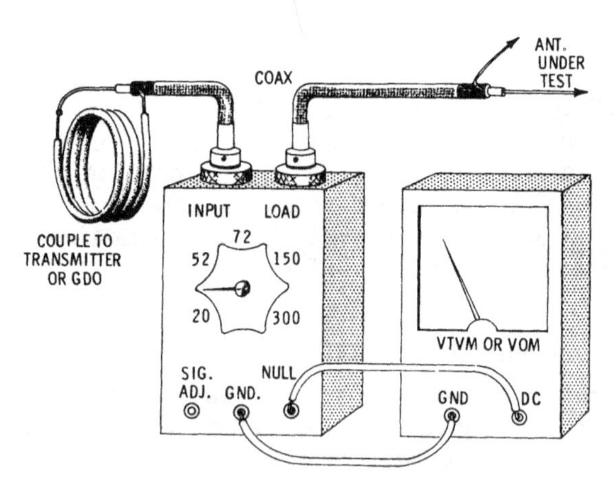

Fig. 3-8. System for measuring impedance.

The exact method of measurement will depend on the instrument in use; however, the following is typical:

1. Power the hook-up by coupling the meter to a transmitter (loose coupling unless power is very low) or to a grid-dip oscillator set to the proper frequency for the antenna under test.
2. Hook the indicating meter (if separate) to the bridge.
3. Increase the coupling to achieve full-scale deflection on the indicating meter.
4. Hook the bridge to the load to be measured.
5. Adjust the control on the bridge for a "null" (minimum reading) on the meter.
6. Read the impedance off the dial.

REFLECTED-POWER SWR METER

This type of meter has wide usage today for a number of reasons: it will handle any power level up to 1 kw; it is easy to hook up; it can be left in the feed line at all times and the SWR reads directly off the dial. This type of meter is available in both kit and manufactured form. Fig. 3-9 shows a typical unit.

Fig. 3-9. A reflected-power SWR meter.

An SWR meter of this type has a variety of uses, most of which stem from the fact that it provides a quick measure of the inductive or capacitive reactance in a circuit. This is most useful data, since, if an antenna is resonant, it will present a purely resistive load to the feeder line of proper impedance when hooked onto it. If it is too long, it will show inductive reactance; and, if too short, capacitive reactance.

Both of these problems can be overcome by proper tuning techniques. For example, a too-short antenna can be loaded up with an inductance until it does present a resistive load. And a too-long antenna can be shortened, in effect, by putting capacity in series with it. Before considering either of these techniques, though, look at the method of achieving critical tune-up of a dipole fed with a coax line, such as the one shown in Fig. 1-4.

Chances are good that if the antenna has been cut to length following the data in Chapter 2, it will offer a satisfactory SWR reading. Actually, you do not need to go overboard in trying to achieve low SWR. The manufacturers of SWR meters uually point out in their literature that variations in coax line impedance, etc. make an SWR reading appreciably below 1.5 to 1.0 as much a matter of luck as anything else. Ordinarily, anything below 2 to 1 is good enough, unless you are using a long coax line at 15 megacycles or higher. However, the situation is the sort of thing where you should do the best you can. And here is the procedure for the tuning coax fed dipole.

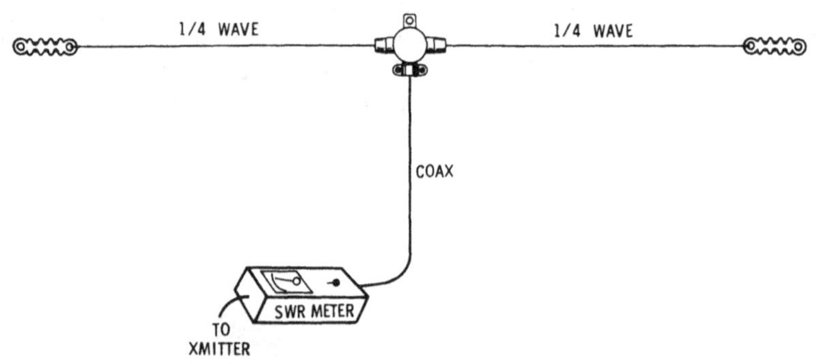

Fig. 3-10. Use of SWR meter to tune up a coaxial dipole.

1. Hook the SWR meter in series with the coax feeder as shown in Fig. 3-10.
2. Hoist the antenna to the height at which it will be operated.
3. Power the antenna with the transmitter.
4. Adjust the SWR meter for a full-scale reading with the function switch thrown to forward power.
5. Turn the function switch to reflected power and read the standing-wave ratio directly off the dial.
6. If the transmitter has a VFO, adjust the transmitter frequency across the ham band in use.
7. You will usually find that the SWR ratio falls to a low value at some segment of the band. For example, it may

be 2 to 1 at 3,600 kc; 1.75 to 1 at 3,700; 1.5 to 1 at 3,800; 1.7 to 1 at 3,900 kc. The frequency range with the lowest SWR indicates the resonant length of the antenna.

If, in the preceding example, you want the antenna to operate its best at 3,800 kc—all well and good. But should maximum efficiency be desired at 3,900 kc, the antenna is a bit long. So remove 6 inches from each end and try again. Perhaps this time you will find that the lowest SWR now occurs at 3,850 kc so you know that by removing only a little more length you will have the ideal length.

Novice-Band Procedure

The preceding kind of tune-up procedure is definitely easier with a VFO. However, if you are a novice, with crystal control, and perhaps only one crystal, you can still do the job.

1. Add a couple of feet to the antenna length as per Table 1-1.
2. Check the SWR reading.
3. Remove 6 inches from each end of the antenna and try again. If the SWR goes down you know you are going in the right direction. If it goes up—add length and try again.
4. Continue to adjust the length until you get the lowest possible SWR reading.

TUNING UP THE GUTTER-SNIPE GROUND PLANE WITH AN SWR METER

By deliberately making an antenna a bit long, and then adding a variable capacitor in series with it for tuning out the inductive reactance, it is possible to provide a simple way to match it to a feedline. Fig. 3-11 shows a hook-up of this type. Notice that the antenna is either .28 wavelength long (for matching to a 52-ohm line) or .32 wavelength long (for matching to a 75-ohm line). The latter is preferable, since it allows a somewhat longer antenna. However, your SWR meter must be capable of handling a 72-ohm line.

For .28 wavelength use the formula $\dfrac{262}{f(mc)}$

For .32 wavelength use the formula $\dfrac{292}{f(mc)}$

The gutter-snipe ground plane described in Chapter 2 uses an electrical hookup which is essentially the same as that in

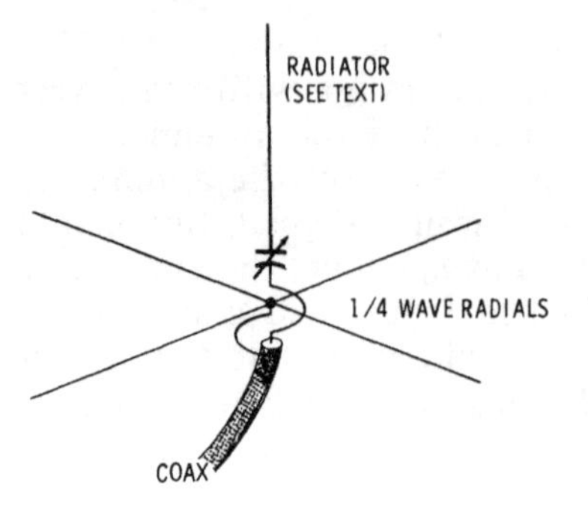

Fig. 3-11. Variable capacitance in series with antenna radiator.

Fig. 3-11. The gutter substitutes for the ground-plane radial. Here is the tune-up procedure:

1. Complete the hook-up as shown in Fig. 3-12. Grounding the gutter at one or both ends may not be necessary, but is a good idea if practical to do.
2. The variable capacitor with powers of 200 watts or less can be a receiving type.
3. Adjust the radiator to the length calculated.
4. Apply power from the transmitter on the desired frequency. Adjust the capacitor for minimum SWR as indicated on the SWR meter.

ADJUSTING AN ANTENNA TUNER WITH AN SWR METER

Fig. 3-13 shows the circuit of an antenna tuner like that described in detail in Chapter 1. The tuner is very easy to

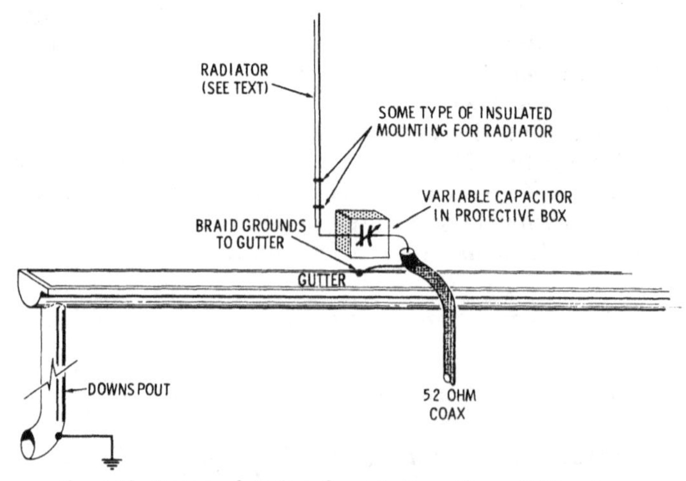

Fig. 3-12. System of tuning the antenna with an SWR meter.

use with the antennas described in the same chapter because the feeder length was chosen to make tune-up simple. However, you may have occasion to feed an antenna with a random length of feeder, in a situation where a long feeder is necessary because of height of antenna, or other consideration.

In such a hook-up, the tapping clips which were provided in the tuner become very important as they allow a transformer action for matching a wide range of impedance. The tune-up procedure is as follows.

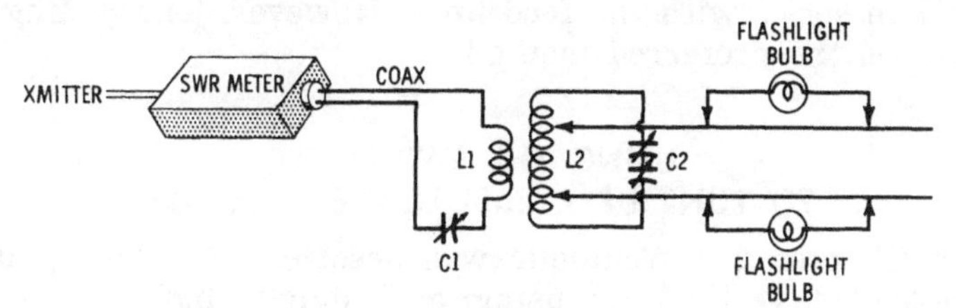

Fig. 3-13. Adjusting an antenna tuner with an SWR meter.

1. Make the hook-up as shown in Fig. 3-13.
2. Place the taps on the ends of coil L2.
3. With the transmitter on, adjust the SWR meter for full-scale reading with the function switch on forward power.
4. Throw the function switch to reflected power.
5. Adjust C2 for minimum reading.
6. Adjust C1 for minimum reading.
7. Readjust C2 for minimum reading.
8. If the SWR is high, try tapping in one turn from the end with each tap. Then go through the foregoing procedure again. If any improvement is noted, move the clips in another turn. The objective is to find the combination of coil value and capacitor which will give the lowest SWR.
9. Always try to keep the taps near the ends of the coil—otherwise coil loss may be high.
10. As an additional check, try tapping a flashlight bulb either in or along both of the feed lines. The bulbs will indicate if the feed-line balance is satisfactory, in the same manner on the RF ammeter described earlier. Of course, an ammeter (or a pair) may be used instead if desired. Also they may show that with the taps close together, the SWR is down—but so is output. If this happens, lengthen the feeder a few feet and repeat the tuning process.

Caution: you may find that even with the taps close together, the SWR ratio as indicated by the meter is quite high. In such a case, the problem is usually one of the feeder being of such a length that it places a current loop at the tuner, and the input resistance becomes extremely low.

Probably the simplest way to solve this problem is to lengthen the feed line by a few feet—the amount needed depending upon the frequency (4′ should do it on 28 mc. On 3.5 mc, perhaps 12′ or more). This may involve rearranging your feed-line layout. If this is not possible, you can accomplish approximately the same thing by adding low-loss coils (identical) in series with the feed lines. However, lengthening the feeder is the preferred method.

USING THE SWR METER TO TUNE UP A BASE-LOADED VERTICAL

In Chapter 2 a technique was described for tuning up a base-loaded vertical by using a flashlight bulb—a method which will give satisfactory results. However, a more accurate match can be achieved by using an SWR meter.

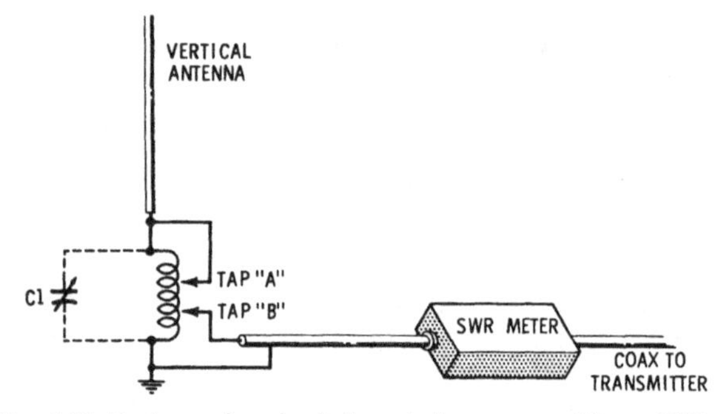

Fig. 3-14. Tuning a base-loaded vertical antenna with an SWR meter.

Fig. 3-14 shows the overall hookup. The tune-up procedure is as follows.

1. Tap the coax feed line on the loading coil, following the suggested starting points in Chapter 2.
2. Hook the transmitter to the feed line. Set the output tuning capacitor for approximately one-half power. Tune the transmitter to resonance.
3. Adjust the sensitivity control on the SWR meter for full-scale reading.
4. Switch to the reflected-power position to read the SWR off dial.

5. Vary tap *A* for lowest SWR.
6. Vary tap *B* to try to achieve a still lower reading.
7. The adjustments are somewhat interrelated—so make small (perhaps fractions of turns) adjustments to achieve lowest possible SWR.

Ordinarily, the above procedure will result in a good match. However, sometimes it is necessary to add capacitance, particularly on 80 and 40 meters. This is capacitor C1 shown by the dotted lines in the drawing Fig. 3-14. Try various settings (100 mmf should be large enough) in connection with adjustments of the other taps to arrive at a combination which gives the lowest SWR. (If used, C1 should be mounted in a watertight box, preferably plastic, to protect it from the adverse weather conditions.)

USING THE GRID-DIP OSCILLATOR

Next to the SWR meter (in the author's opinion at least) the most useful piece of antenna test equipment is the grid-dip oscillator (Fig. 3-15). Further, it is an extremely useful device for many other tune-up and design jobs.

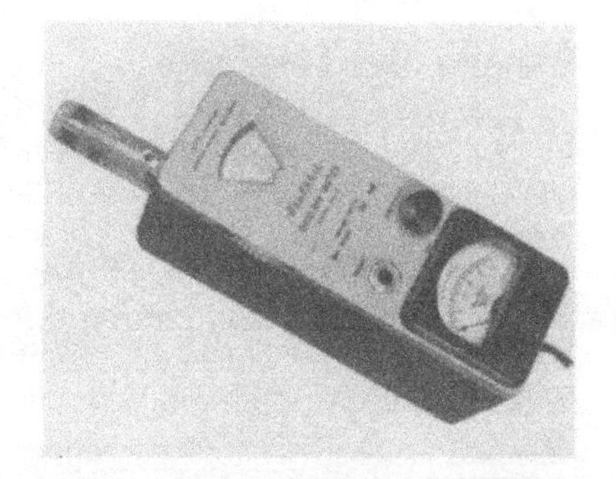

Fig. 3-15. A typical GDO meter.

Actually, a grid-dip oscillator is a high-frequency oscillator, usually covering a range of 2 to 250 mc or more. A sensitive microammeter is inserted in the circuit (usually in the grid circuit) to measure current. When the tuning coil of the oscillator is coupled to a resonant circuit (or antennas), sufficient loading takes place so that there is a very sharp dip in current (as indicated on the meter). Hence, the name of the device—grid dip.

Uses of the grid-dip oscillator are legion—only a couple of the more frequent applications will be covered here.

Achieving Antenna Resonance

Fig. 3-14 shows the circuit of the vertical antenna discussed previously. Initial placement of tap *A* can be achieved very easily if a GDO is available.

1. With your receiver and BFO on, tune the receiver to the frequency to which you would like to tune your antenna.
2. Turn on the GDO, and, with the proper coil plugged in, hold the GDO close to the antenna lead-in for the receiver.
3. Adjust the GDO for zero beat with the receiver. This is indicated as you turn the GDO dial by (a) a whistle in the receiver speaker, (b) a quiet area, (c) then, as you advance the dial, a whistle again. The zero beat is the point on the dial between the whistles—and is usually quite critical.
4. Now, keeping the same dial setting on the GDO, couple the GDO loosely to the loading coil at the base of the antenna.
5. Adjust tap *A* until you get a sharp dip on the GDO indicating meter. When this occurs, you know you have tuned the antenna to the same frequency as that on your receiver.

Checking Out Tuner Coils

The data in Chapter 1 tells in detail how to build a tuner of the type shown in Fig. 3-13. However, you may have occasion to depart from the exact design, perhaps to use some coils which are already in your junk box.

With a GDO, it is a simple matter to check the resonant frequency of an LC combination. For example, suppose you want to find out if the LC combination you have is hitting 28 megacycles, as planned.

1. Set the GDO to the desired frequency, either by referring to its dial or, preferably, by beating it against the receiver BFO as described previously.
2. Couple the GDO to the coil in the tuner.
3. Rotate the variable capacitor on the tuner. If the coil-capacitor combinations are right, you will bet the indicating dip well within the tuning range of the tuner capacitor.
4. If the dip does not occur, you can rotate the GDO dial until you do find it. If the dip point is considerably lower in frequency than desired, the next step, of course, is to

reduce the number of turns on the coil. If it is higher, increase the turns.

The same method, of course, can be used to determine whether a capacitor and coil are tuning to the desired frequency in a transmitter or receiver.

Adjusting a Phasing Line

In the vertical-beam antennas (described in Chapter 5) considerable use is made of lengths of coaxial cable cut to the proper dimensions to provide the matching and phasing effects necessary to create beams.

These phasing sections can be made simply by referring to the data in the chapter, or by simple calculations. However, for the most accurate possible lengths, the GDO can be used to determine proper dimensions. Here's how the job is done:

1. The usual objective is to have a line one-quarter wavelength long. Therefore the first step is to cut the line to the length plus 10%.

Fig. 3-16. Part of a hookup for adjusting a phasing line with a GDO.

2. Fit one end of the line with a coaxial plug. Insert this plug into a coax receptacle which has a single turn of wire attached to it, as shown in Fig. 3-16.
3. Couple the grid-dip oscillator to this coupling loop.
4. With the GDO dial, determine the resonant frequency, cut off the line an inch at a time until you find the point at which the line is resonant. The line should then be fitted with a connecter plug.

Using the GDO with a Bridge-Type SWR Meter

Many hams either have simple bridge-type SWR meters available, or build them up from one of the amateur handbooks or magazines. The meter is comprised of a circuit like that shown in Fig. 3-17. Such an SWR meter only indicates minimum SWR. However, this is adequate for many purposes. A serious difficulty with such meters, though, is that they are very easily damaged by being supplied too much power.

One good solution to this is to use the grid-dip oscillator as the powering source since the output is naturally quite small. This does assume that the GDO will stay on frequency once it is set up. Because coupling to the grid-dip oscillator as a source of power usually throws it off frequency, the frequency should be checked against a receiver using the technique outlined previously. And, be certain that the whole setup is mechanically solid enough so that the grid-dip oscillator is not detuned as the result of the cable moving in relationship to it. That is, the dial being bumped accidentally, etc.

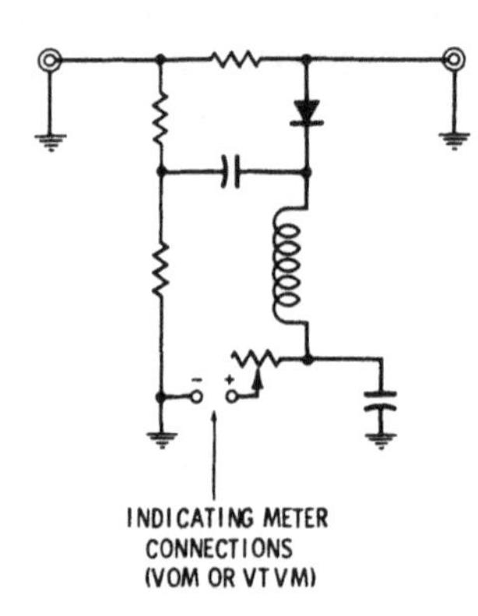

Fig. 3-17. Using the grid-dip oscillator with a bridge-type SWR meter.

Resonating a Dipole

In Chapter 2, one of the antennas described was the 33-Out 23-Up. This antenna can be tuned for resonance with a pick-up bulb by using a SWR meter or with a grid-dip oscillator. With the latter method, the GDO can be set on the desired frequency, coupled to the loading coil, and then the coil adjusted by tapping it until the desired grid dip is obtained on the indicating meter. Here, this technique is easy because there is a coil to which we can couple. The job becomes a bit more difficult when the dipole is hoisted into the air. Perhaps the best way to do the job is to provide some temporary mount for the antenna about 10 feet in the air; and then following Fig. 3-18 check the frequency with the GDO.

Another way to accomplish the matter is to hook temporarily on a half-wave length (or some multiple of a half wave) of 72-ohm twin-lead feeder line. The reason for using this is that assuming the feed line is of the proper length, coupling to the line is the same thing as coupling to the center of the antenna.

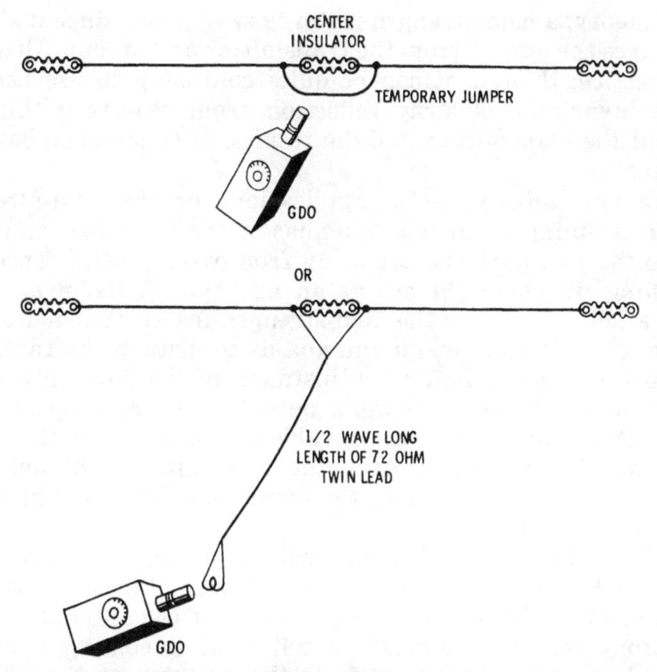

Fig. 3-18. Determining the resonant length of a dipole with a GDO.

Fig. 3-18 shows the hook-up. The frequency of the dipole can be read directly off the dial of the GDO.

USING A FIELD-STRENGTH METER

Fig. 3-19 shows a typical field-strength meter. This particular meter has a transistorized amplifier which increases the meter sensitivity and improves the accuracy of measurement over that of a simple diode-type meter.

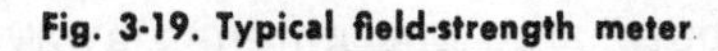

Fig. 3-19. Typical field-strength meter.

In theory, a field-strength meter is easy to use since it simply measures the output from the transmitter and antenna directly. In practice, though, it can be quite confusing to use because there invariably is stray reflection from objects within the area of the transmitter, and the results, of course, can be quite misleading.

One particularly useful application for the field-strength meter is tuning-up mobile antennas of the type used on autos. To do the job, park the car away from wires, metal fences, or anything which might act as an antenna on its own. Then, with a helper reading the field-strength meter (but not standing so close to the pickup antenna as to disturb the tuning of the meter) make whatever adjustment of the auto antenna as is necessary to give maximum output. The accuracy of field-strength measurements are invariably improved if the meter is a fair distance away from the transmitting antenna. For this reason, always work at the maximum distance which will give a satisfactory reading.

Using a field-strength meter with a vertical antenna which is grounded for the actual earth is a very difficult procedure because ground currents in the vicinity of the antenna may be so strong that it is difficult to tell what is coming from the ground and what is coming from the antenna itself.

A field-strength meter can be used with average results in the tune-up of a parasitic-type beam. The beam should be such that it can be rotated and yet be reached conveniently. One way to do the job is to adjust the elements for maximum output starting with the radiator itself; with both the reflector and the director deliberately adjusted to a dimension considerably less than that of the radiator. Once the radiator is adjusted to the maximum output, the next step is to adjust the forward, director element. Finally, adjust the length of the back, reflector element.

While the preceding method will result in maximum output, it seldom will achieve maximum front to back ratio, which is the ability of the antenna to receive signals from the front while, at the same time, reducing signals coming in from the opposite direction. To achieve this, follow the same general procedure as outlined above. After the radiator has been tuned for resonance and maximum output, point the antenna away from the field-strength meter and first adjust the reflector, and then the director to give the minimum meter reading.

If you have a patient ham friend who lives nearby, you can accomplish the same thing as using a field-strength meter by having him report on signal strength from your rig as shown on the *S* meter on his receiver. Actually, in some respects, this method is even more accurate, since it tends to eliminate some of the effects of reflections from nearby objects. The technique is particularly useful for determining the front-to-back ratio of your antenna.

CHAPTER 4

Multiple-Element Beam Antennas

Once they have a basic antenna for several bands, most hams want some type of beam, particularly for 14, 21, and 28 mc. The lucky ones, those who can afford several hundred dollars for a commercial beam-trap antenna complete with tower and heavy-duty rotator, can get the job done without much effort. However, this book is intended for the younger hams, or those with thin purses, who don't have too many dollars to spend on their hobby. Fortunately, there are a number of ways to build effective beams for surprisingly little money. But, before working out some specific antennas, there are some things you should know about beams.

BASIC TYPES OF MULTIPLE-ELEMENT BEAMS

In the case of multiple-element beam antennas, there are four basic types: end-fire arrays, broadside arrays, collinear arrays, and parasitic arrays. Tuning up beams and pointing them in the correct direction will be greatly simplified if you understand how each of these four types work. First we will consider the driven array, so-called because power is applied to each of the elements.

End-Fire Arrays

The 8JK to be described in detail later in this chapter is an end-fire array. Essentially, it is made up of two dipoles, fed 180° out of phase. The overall theory behind the functioning of such beams will be left for more technical books. Suffice it to say that when two dipoles are fed 180° out of phase, and are properly spaced, their output adds in the favored directions and at low angles. This increases power greatly. For example, it is easy to achieve a 3-db gain which is equivalent to doubling the power output.

Fig. 4-1 shows the output pattern as compared with a dipole. Note that the pattern, instead of covering 360°, has two narrow lobes. This narrowing is where the extra power comes from. Further, the radiation is concentrated at low angles, the useful angles on 14 mc and higher.

While Fig. 4-1 deals with radiators mounted parallel to the ground, the same principle applies when the radiators are mounted vertically. In this case, imagine you are looking down on the end of the vertical radiator and that the pattern is the pattern on the earth, the dotted line, of course, is the theoretical pattern from a single vertical element.

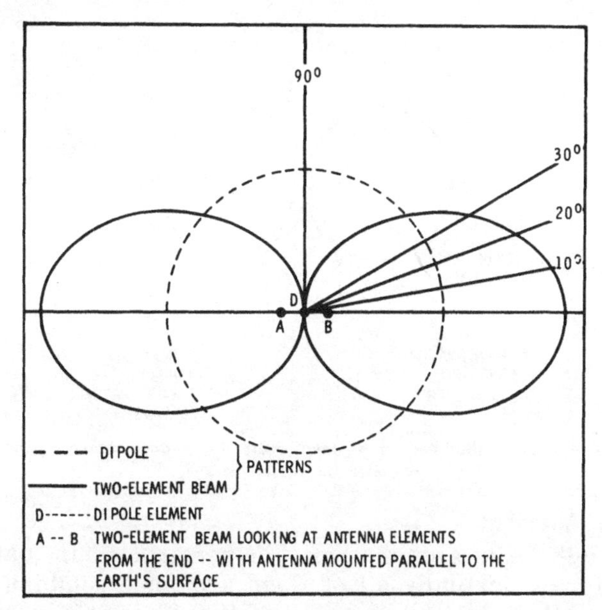

Fig. 4-1. How a beam antenna creates a power gain by narrowing the radiation pattern.

Broadside Arrays

By spreading out the dipoles a bit further and feeding them in phase instead of 180° out of phase, we create a broadside array with a different pattern. If the dipoles were parallel to the ground, the pattern actually would not be worth much, since the antenna would transmit straight up and down. However, by mounting the radiators either on end, or one above the other, we get a useful directive pattern. Fig. 4-2 shows the pattern from both a broadside and an end-fire array with the radiators mounted vertically.

Note that with the radiators in the vertical position, the patterns of both broadside and end-fire arrays are at right angles to each other. This difference can be achieved by feeding the antennas either 180° out of phase (end-fire) or in phase (broadside). Thus, if we choose a compromise spacing for the radiators (3/8 wavelength is ideal), and have some simple way of changing the length of the feedline between the radiators, we can shift direction of the beam simply by adjusting the feeder length. This is a most useful technique, and one which makes possible some highly effective vertical beams to be described later.

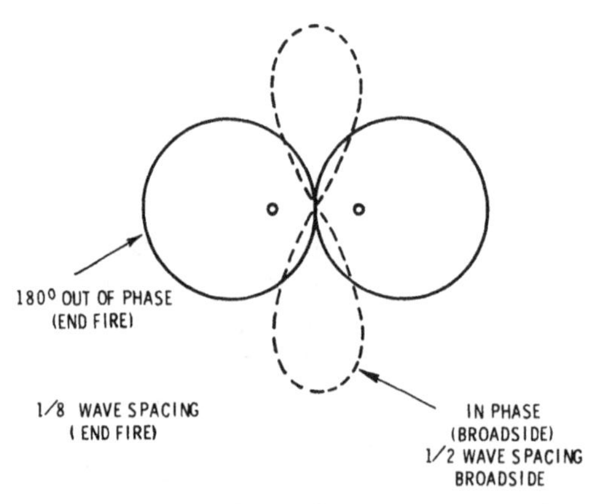

Fig. 4-2. How the method of feed shifts the pattern direction on a two-element driven array.

Collinear Antennas

Although beam antennas may be constructed in many different forms, virtually all of them are some combination of end-fire or broadside arrays (or of the other two basic beam types—collinear and parasitic arrays). When a 40-meter, center-fed antenna is operated on 20 meters, it is actually a collinear array. The output from the two radiating sections combine (the pattern narrows down), and the result is some gain (1.9 db) in the favored direction (Fig. 4-3). Such an antenna is often called "two half waves in phase."

By stacking two collinear arrays one above the other, and thus creating, in effect, a broadside array as well as a collinear array, we create an effective beam called a Lazy H. Fig. 4-4 shows such an antenna. This antenna can be fed with a tuned line and the antenna tuner described in Chapter 1. The horizontal-directive pattern will be the same as for the collinear

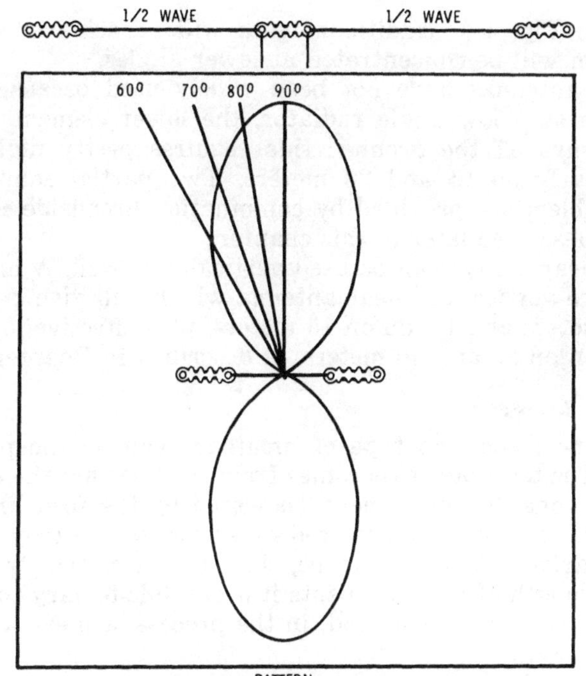

PATTERN
LOOKING DOWN ON ANTENNA WHICH IS HORIZONTAL TO THE EARTH
NOTE PATTERN IS APPROXIMATELY 60° WIDE AT 1/2 POWER POINT

Fig. 4-3. Radiation pattern of a collinear array.

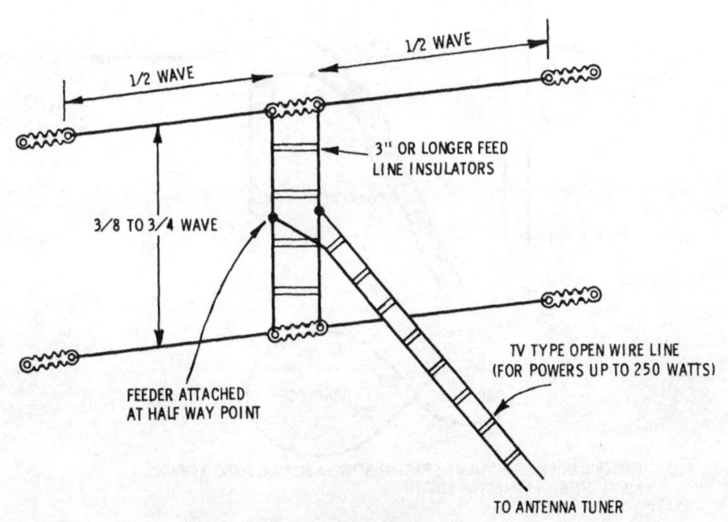

Fig. 4-4. Lazy-H antenna.

array in Fig. 4-3. Additional gain will be achieved because radiation will be concentrated at lower angles.

Such antennas have not been widely used because, to be effective as a low angle radiator, the lower elements should be ½ wave off the ground; this requires pretty high poles, particularly on 15 and 20 meters. Two partial solutions to this problem are provided by combination broadside-collinear arrays described later in this chapter.

Collinear arrays can be used vertically as well. A high performance vertical-collinear antenna which will give 3-db gain on 10 meters and 1.9 db on 15 meters, plus effective low-angle radiation on 20 and 40 meters, is described in Chapter 5.

Parasitic Arrays

The most common type of amateur beam is the parasitic array. The term *parasitic* comes from the fact that the antenna has only one driven element (powered by the feed line) and the other elements are powered by energy which they pick up off the driven elements. Hence, they are parasitic. By adjusting the length of these elements it is possible to vary the phase of the current on them; and, in the process, achieve a variety

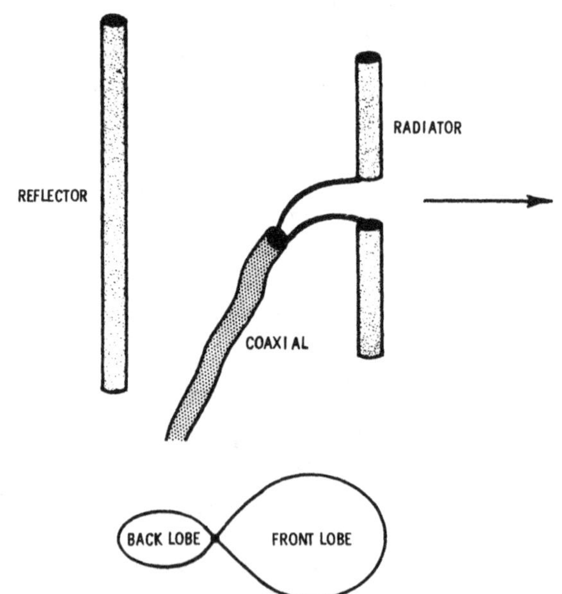

Fig. 4-5. Two-Element Parasitic Array using reflector.

of effects which result in a multiplication of power from the antenna. For example, by making the parasitic element longer than the radiator, the parasitic element becomes a reflector, increasing the signal in the direction of the radiator (Fig. 4-5). The gain is 3 to 4 db over a single element. By making the parasitic element shorter, it becomes a director, and the signal is again stronger away from the radiator (Fig. 4-6). Again the gain is 3 to 4 db.

Oftentimes, a reflector and a radiator are combined with a radiator to create a three-element parasitic array capable of 6 db or more of gain (Fig. 4-7). Such antennas are very popu-

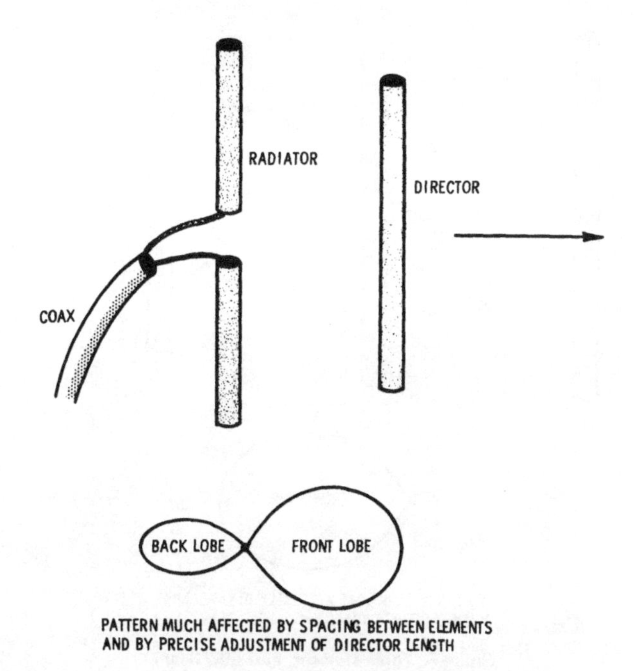

Fig. 4-6. Two-Element Parasitic Array using director.

lar, particularly on 14 mc and higher; since, at this frequency, they become small enough to be rotated. Likewise, as vertical radiators, two elements can be combined to form a beam, the direction of which can be changed simply by lengthening or shortening one element. Vertical parasitic arrays have not seen wide usage, though, because interference from nearby objects often upsets the directional pattern unless the beam is mounted fairly high. For the vertical beam mounted on the ground, the driven array is better.

WHAT DIRECTION SHOULD A BEAM POINT?

Because beam antennas achieve their increased power in favored directions by narrowing the output pattern, it is most important that the beam be aimed in the right direction. This need for *aiming* has led to the development of rotating beam antennas, which is an excellent solution, although somewhat expensive.

Fortunately, for the ham on a small budget, a beam with a bidirectional *figure-8* pattern will do a fairly good job of covering the United States from most locations. In addition,

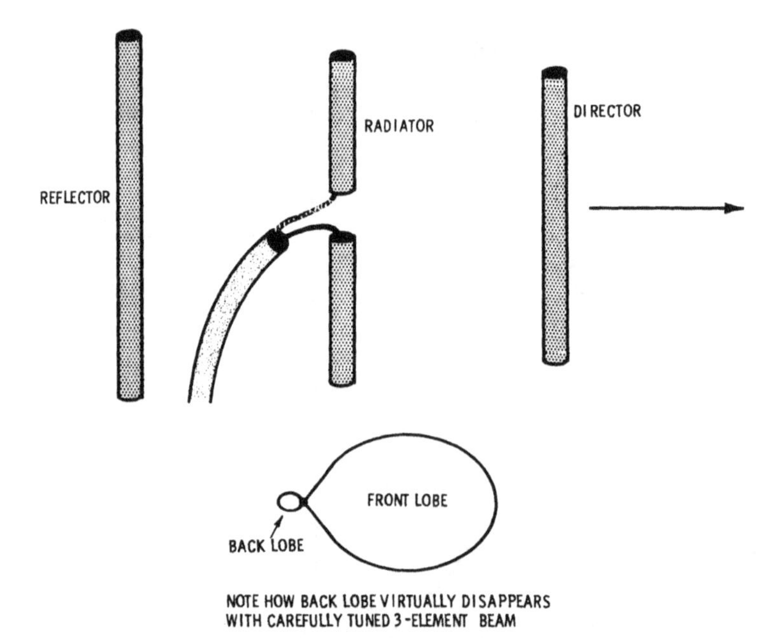

Fig. 4-7. Three-Element Parasitic Beam.

most DX countries which are well populated by hams lie along two lines so that even a *figure-8* pattern beam will reach a lot of countries. If you have a switching arrangement which provides two *figure-8* patterns at right angles to each other, you'll get good 360° coverage.

Perhaps all this will be clearer by referring to Fig. 4-8. This is a world compass, worked out with its center on Wichita, Kansas. Notice how a *figure-8* antenna pattern covers Europe *plus* New Zealand and Australia. By adding another *figure-8* pattern we can pick up most of the rest of the world.

To help you lay out your antenna system so that you can get the directions you want, world compasses are provided for three cities: San Francisco, Wichita, and Washington, D.C. Figs. 4-9, 4-10 and 4-11 respectively. Pick the city closest to you; then you can use a simple protractor to get the other directions and aim your antenna with sufficient accuracy for all practical purposes with a simple beam.

In many cities, streets are layed out North to South and East to West, so determining direction isn't too hard. But the job is not so easy if you live on a curved street, or streets set at an angle to the basic directions. Probably the best way to

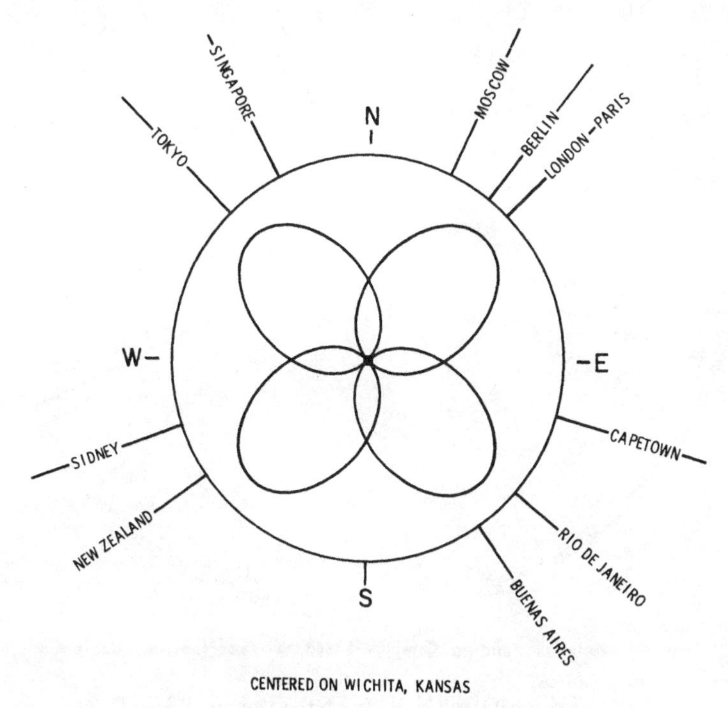

Fig. 4-8. How two Figure-8 patterns will cover most of the world.

determine true North is by means of the North Star. Any high school astronomy text, Boy Scout handbook, etc. will tell you how to find the star (also called Polaris).

Using Polaris as a way to determine North means that you won't have to worry about compass corrections, as will be the case with an ordinary compass. *If* you use a compass, however, phone the local airport control tower (or city engineer's office) to find out how many degrees of compass correction is necessary. Remember that the correction for North is *opposite* the number of degrees or compass variation. For example, a correction of 10° East means that true North is 10° to the West of the compass needle heading.

Know which direction to head your antenna? Fine—now let's take a look at a simple, practical beam antenna.

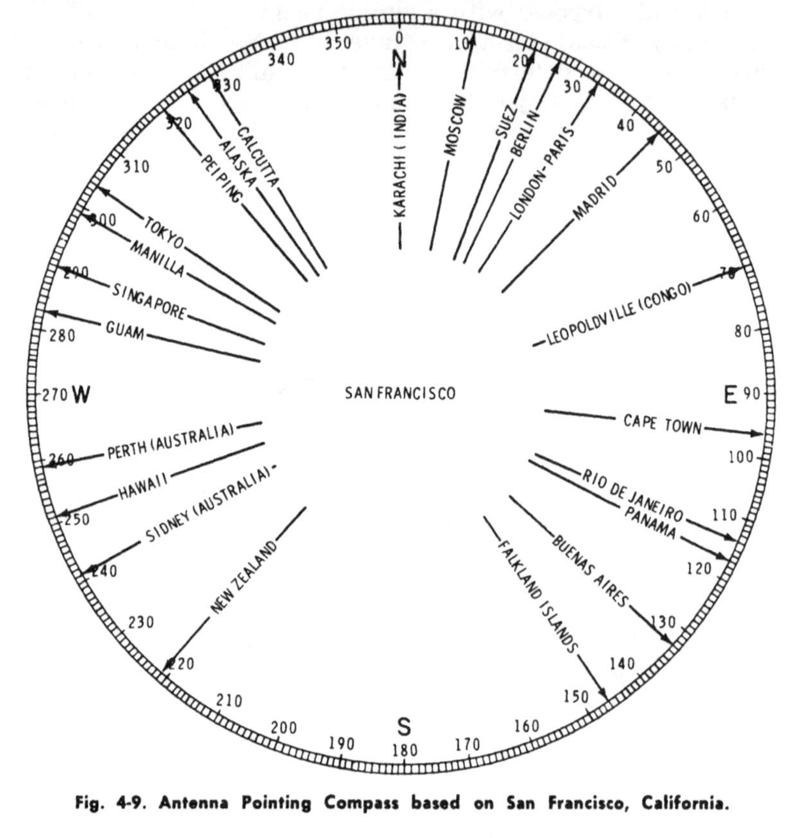

Fig. 4-9. Antenna Pointing Compass based on San Francisco, California.

THE MODERN 8JK END-FIRE ARRAY

For quite a number of years, an extremely popular antenna was the 8JK, named after the ham call letters of the ham who first popularized it. It provided good gain, low-angle radiation, and was suitable for home construction. On the negative side of the ledger, it was quite sharp in tuning, did not lend itself to the all metal type of construction which eventually became popular, and it was bidirectional. The last named feature, which can be an advantage as well as a disadvantage, will be discussed later.

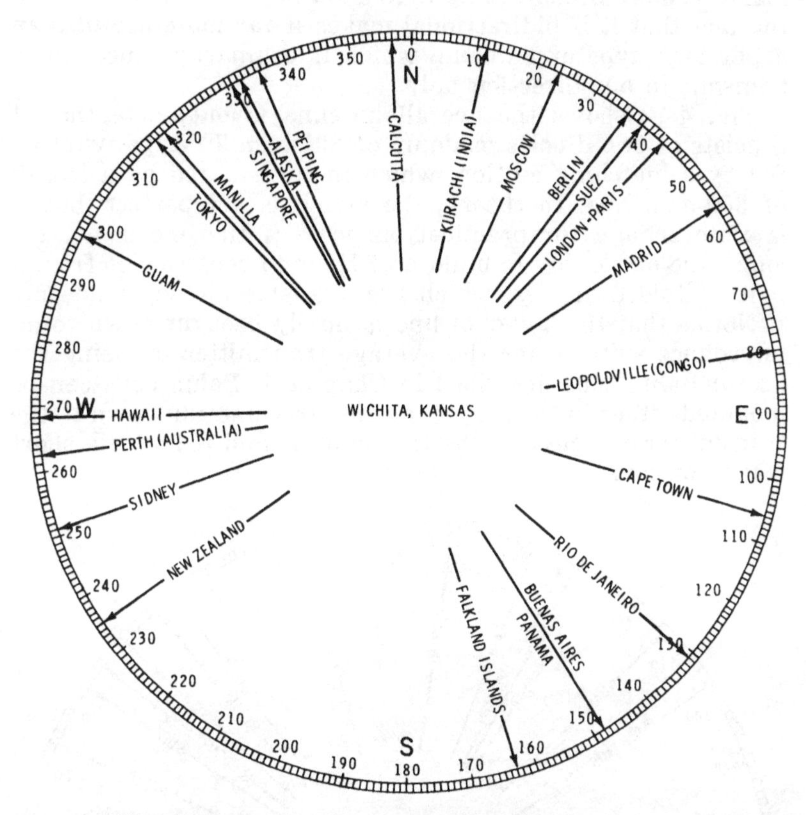

Fig. 4-10. Antenna Pointing Compass based on Wichita, Kansas.

Improved materials and techniques have made the negatives less important. In fact, as a first beam for a ham to build, the 8JK has much to recommend it.

An Indoor Beam

The 8JK, for example, is an ideal antenna for an indoor installation. An indoor beam will usually "run the socks off" any simple dipole mounted outdoors at the same height. Using an 8JK identical to the one to be described, the author contacted stations as far away as Kwajelin (from Denver) with a 50-watt phone rig. The antenna was in the attic near a maze of electrical conduit, plumbing pipes, heating runs, etc.—a lot too close for comfort.

The 8JK is ideal for an indoor installation for two reasons. It can be constructed easily of ordinary TV twinlead (which makes it more broad-tuning than a simple wire antenna). And the fact that it is bidirectional makes it far more useful than a parasitic type of antenna, which is ordinarily tuned up to transmit in one direction only.

Fig. 4-12 shows the overall antenna layout; note that it consists of two dipoles made up of 300-ohm TV-type twinlead fed by a matching section, which in turn, is fed by a length of 300-ohm line. In theory, the match is not perfect, but it is close enough for practical purposes if the line is not too long. One quick way to build an 8JK is to construct it from a pair of folded dipoles, which are available already made up.

Notice that the 300-ohm line is finally brought down to an impedance suitable for the average transmitter by using the useful Balun coils described in Chapter 1. Balun coils can be mounted either in the attic (and the coax line run down to the transmitter) or close to the transmitter, and fed with a short length of coax.

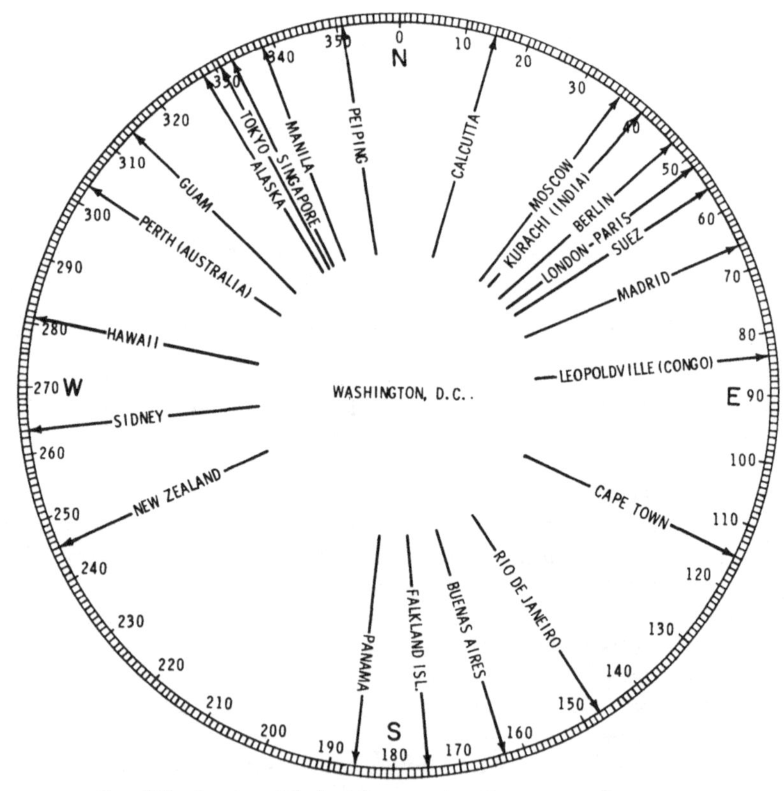

Fig. 4-11. Antenna Pointing Compass based on Washington, D.C.

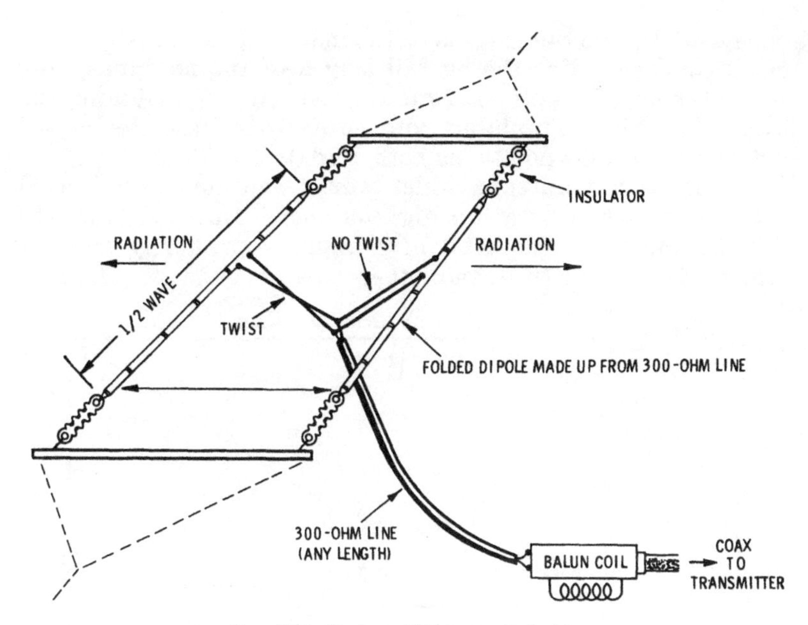

Fig. 4-12. Modern 8JK (gain 4 db+).

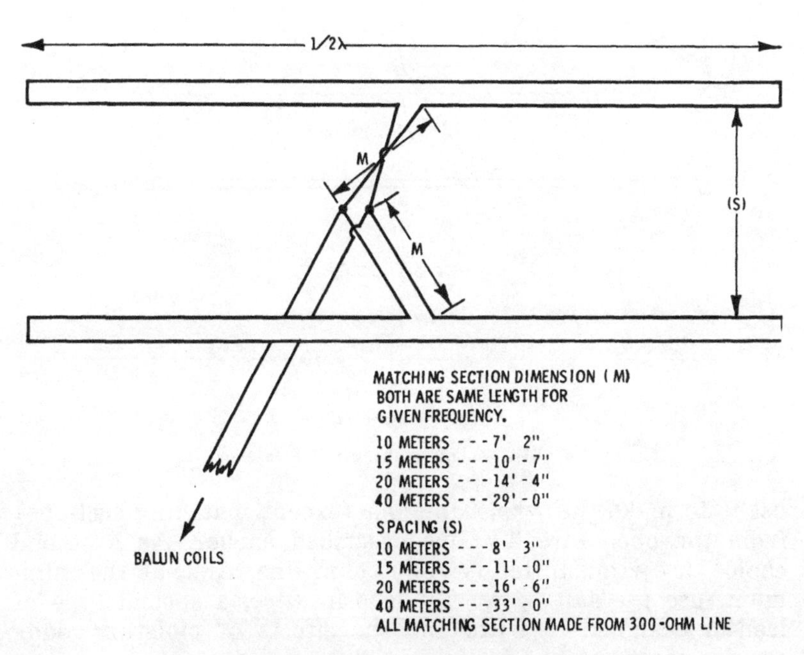

Fig. 4-13. Indoor beam matching-section dimensions.

Fig. 4-13 gives the lengths of the matching sections for various frequencies. For the overall length of the antennas refer to the tables in Chapter 1. Note the twist in one matching section lead; this is absolutely vital, otherwise the antenna will not work (but do not twist both leads).

Of course, any antenna which works well indoors will work even better when hung up high outdoors. However, the 8JK for outdoor use needs a bit of changing—primarily to avoid trouble from moisture and other weather effects. The best

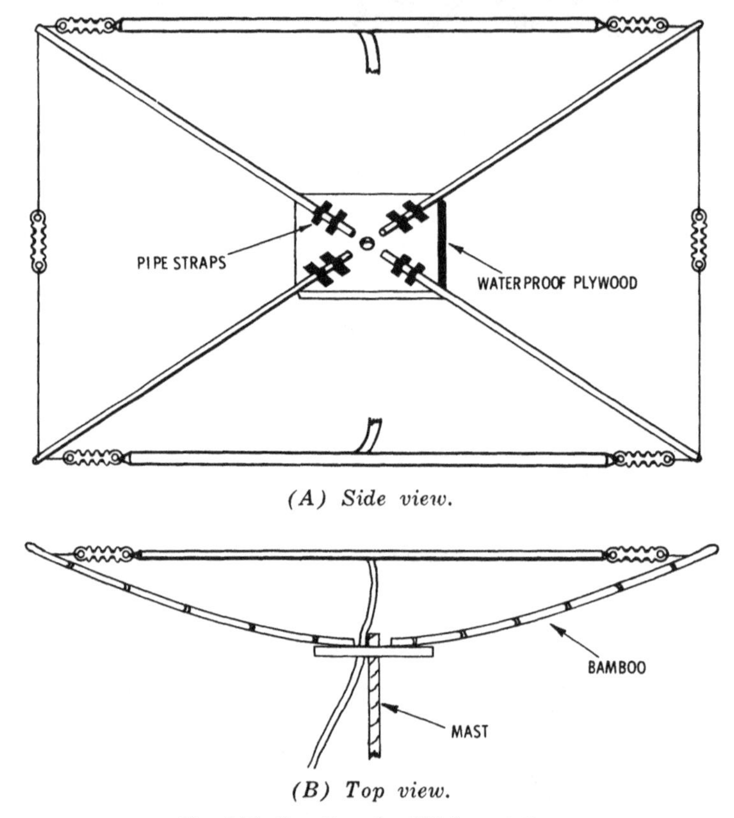

(A) Side view.

(B) Top view.

Fig. 4-14. Mounting the 8JK for rotation.

bet is to make the whole antenna (except matching sections) from the open-wire TV line described earlier. As a second choice, use transmitter type 300-ohm line. And, at the minimum, use the salt spray TV lead-in wire, a special type of lead-in designed to withstand the effects of moisture along the sea coast.

The antenna segments can be mounted with a spreader hung from poles (Fig. 4-12) or on center-mounted spreaders as shown in Fig. 4-14. Bamboo is ideal for the latter purpose; to reduce weathering, it should be given a couple of coats of spar varnish. A 10-meter beam made on the spreaders can be rotated easily from the ground; see Chapter 6 for one method.

TWO BROADSIDE ARRAYS

As mentioned earlier, the *lazy-H* antenna has not been used to the extent that it deserves, considering the fact that it will give a gain of as much as 6 db (equivalent to increasing power

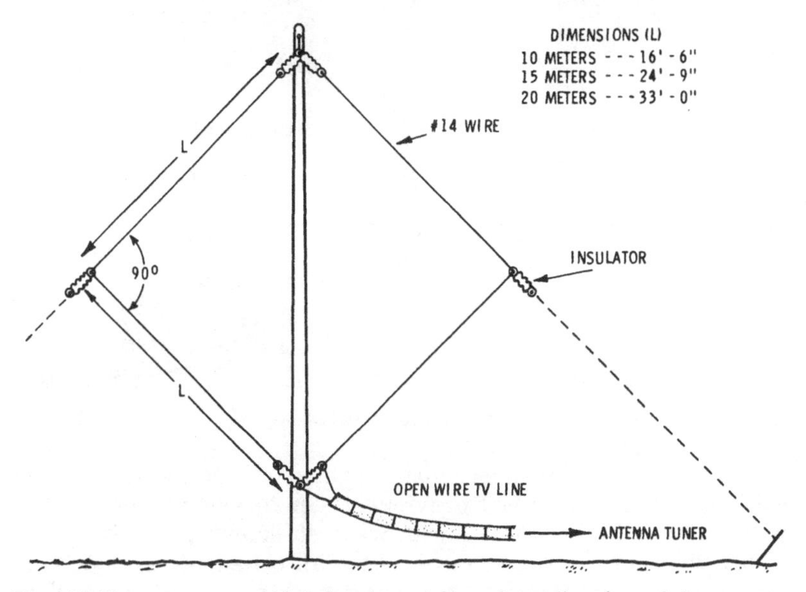

Fig. 4-15. Bisquare antenna with 4-db gain at right angles to the plane of the antenna.

4 times) without excessively sharp tuning and with the possibility of multiple-band operation. The problem, of course, is usually one of getting poles high enough to support the antenna.

Two modifications of the *lazy-H* solve this problem to some extent. The first of these is the Bisquare, which was developed by hams on the West Coast. The bisquare looks like a vertically-polarized antenna, but actually it is horizontally polarized on the prime frequency. It requires only one pole, and the elements are bent, saving overall height. Fig. 4-15 shows the overall layout. Get the bottom feed point as high off the ground as possible—at least 4 feet for 28 mc, and even higher for 14 and 21 mc. The antenna can be fed most easily with a tuned feed line off the tuner (described in Chapter 1).

Another variation of the lazy-H was worked out by an Australian ham (VK2SA) for use on 40-20-15. While the theory in back of this particular antenna is somewhat baffling, it works well in actual practice, and the bottom elements make an important contribution to the antenna effectiveness (Fig. 4-16).

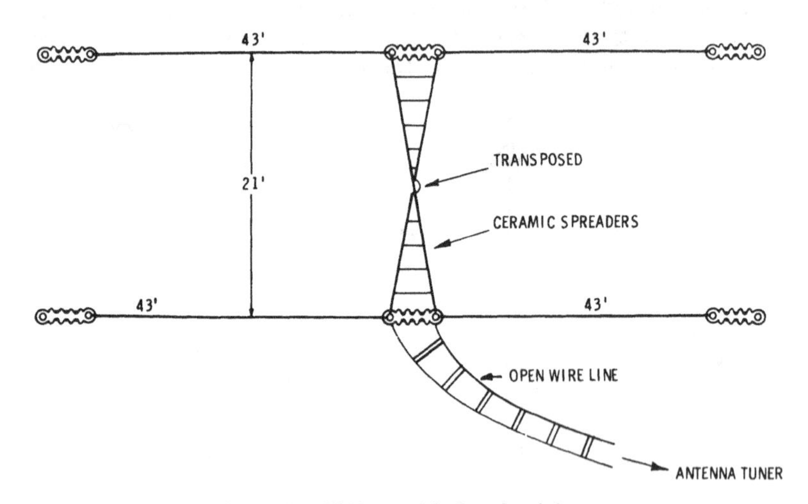

Fig. 4-16. VK2SA special three-band beam.

On 40 meters the directivity pattern will be similar to the collinear array described previously. On 20 and 15 meters, the pattern will be multiple-lobed with sufficient lobes to cover fairly well around the compass. Like all broadside arrays, this one needs all the height you can get, although the relatively close spacing between the radiating elements makes the problem less acute than in the case with the standard lazy-H.

PARASITIC BEAMS

These days, building full-sized, low-frequency parasitic beams is usually more trouble and expense than it is worth, especially when good-value commercial beams of this type are readily available. For that reason, none are described in this book. However, later chapters do include descriptions of a low-cost, miniature beam and describe home-brewed VHF beams which can be built more easily.

Driven arrays lend themselves more to home construction, particularly vertical beams which can be built quickly and easily by combining low-cost vertical antennas of the type currently being manufactured by at least three companies. Chapter 5 describes such antennas in detail.

CHAPTER 5

Vertical Beams

As mentioned previously, the parasitic type of horizontal beam has long been popular. It is fairly compact, can be rotated, and is easy to feed. These advantages are somewhat negated by its high cost (particularly on 14 mc and lower), the need for considerable height for low-angle radiation (60 feet is not a bit too much on 14 mc), and its poor appearance when mounted in a residential area. Further, the big horizontal parasitic beam does not lend itself well to home construction. So, if you want one, better plan on a commercial unit.

Vertical beams, on the other hand, are low in cost, quick and easy to build, can be rotated electrically, do not clutter up the landscape too badly; and all tuning adjustments are located on the ground where they are easy to get at. On the negative side, vertical beams may waste power through absorption by nearby objects. They also require a good ground, particularly on 80 and 40 meters, and tend to pick up ignition noise on 14-21-28. But, on balance, they have a lot to recommend them.

For 80 and 40 meters, the vertical beams have another advantage in that proper spacing is practical. Proper spacing is almost impossible with a mechanically-rotated array on these frequencies. As one example, one-eighth wave spacing is about the minimum which can be used between elements on any type of array without running into very low radiation resistance, high losses, and extremely sharp tuning, (which, in turn, restricts operation to a very small portion of an amateur band). Yet, on 80 meters, one-eighth wave spacing means that the elements must be 33 feet apart. In turn, this requires a 33-foot boom for even a 2-element rotary beam, and a 3-element beam would require a 60-foot boom. Such structures are impractical for most hams.

With vertical beams, though, these spacings are practical. For example, placing two vertical elements 33-feet apart at the rear of a lot is perfectly feasible for most hams who live in houses, even if the lot is as narrow as 50 feet. By running the verticals down the side of the lot, wider spacing is practical. And, as the antenna descriptions in this chapter show, 40-, 20-, 15-, and 10-meter antennas each require progressively less space.

So much for the inherent advantages and disadvantages of vertical beams. Now let's take a look at some practical designs.

A VERTICAL BEAM FOR 10 TO 15 MC

By making an antenna one full wavelength long and feeding it in the center, it is possible to achieve a definite beam effect. Fig. 5-1 shows the pattern of such an antenna as compared

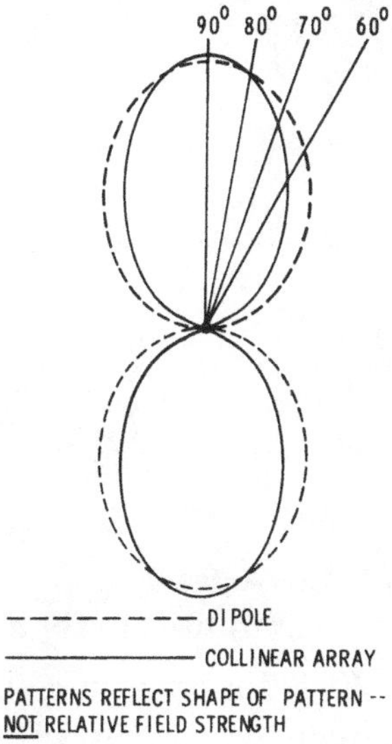

Fig. 5-1. Comparison of dipole and collinear array patterns.

with a dipole. Such an antenna is called a collinear array. If the length is increased still more, the pattern becomes even narrower (Fig. 5-2) and the gain goes up. Although such antennas are sometimes used horizontally, they can also be mounted vertically. They will then give a 360° pattern with the signal at a very low angle, the kind which produces real DX.

These antennas have not been widely used, because even on 10 meters they are pretty high in the air. However, by using a commercial 23-foot vertical antenna and mounting it on the side of the house or on a pole and then using a loading coil to piece out the lower leg, the whole idea becomes quite practical.

Such an arrangement is not perfect (for example, you won't be able to obtain a perfect feeder-line balance) but it will work out well, and it will yield a power gain approaching 3 db—which is equivalent to doubling the output power, Actually, it will do more than that in effective DX work, since all radiation is concentrated at an extremely low angle.

Such an antenna is best fed with a tuned line, since the feed point is of high impedance. This also makes it possible to use the antenna on 21 and 14 mc as well, although the gain on 21 is in the 1.5-db range, and negligible on 14. However, even on 14 mc, the antenna shown in Fig. 5-3 will yield the low-angle radiation needed for DX.

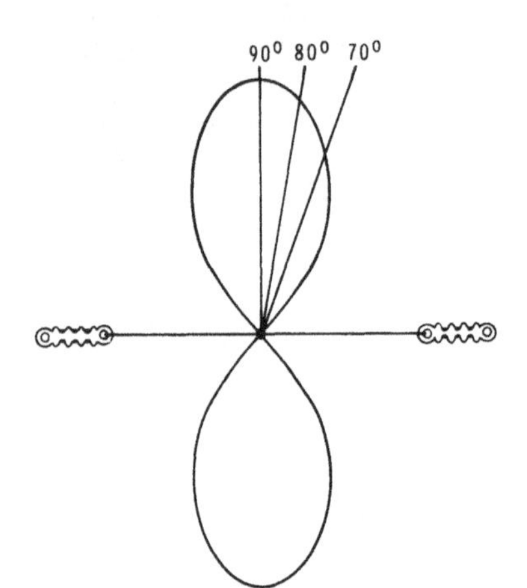

Fig. 5-2. Radiation pattern of collinear array with lengthened element (extended double Zepp).

Tuning The Antenna

It is best to tune first on 15 meters.

1. With the proper coil in the tuner, adjust the variable capacitor in the tuner to resonance as indicated by a flashlight bulb and loop coupled to the tuner coil.
2. Adjust the tap on coil L1 for maximum output as indicated by the pick-up loop coupled to the coil.

Ordinarily, once the antenna has been adjusted to resonance on 21 mc, it will tune easily on both 28 and 14 mc simply by using the proper coil in the tuner.

THE 50-MINUTE-SPECIAL, 10-METER ARRAY

Here is a beam so simple that you can actually erect it and have it on the air in less than an hour, particularly if you build it from two commercial 23-foot all-band verticals. If you make it from scratch (using two radiators like the home-made all-band vertical described in Chapter 3) the job is still one you can polish off in a day.

The antenna is a bidirectional, end-fire array, using two ¾ wave radiators, and has the *figure-8* radiation pattern shown in Fig. 5-4. The radiators, simply because of their length, de-

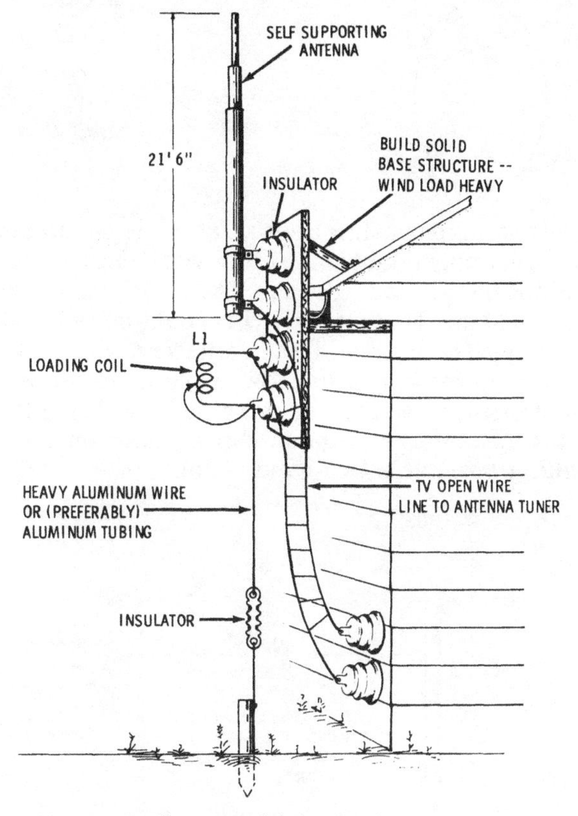

Fig. 5-3. Simplest vertical beam for 10 to 15 meters.

velop low-angle radiation. And when two are combined, the low-angle effect is enhanced. The result is an antenna with real possibilities for DX. It is no slouch on short range 2000-mile contacts either. For the author, the test antenna yielded a "10 db over 9" report from the first station called, which wasn't bad at all considering that the rig in use at that time was a little 30-watt phone transmitter. Fig. 5-5 shows the overall layout of the 50-Minute-Special.

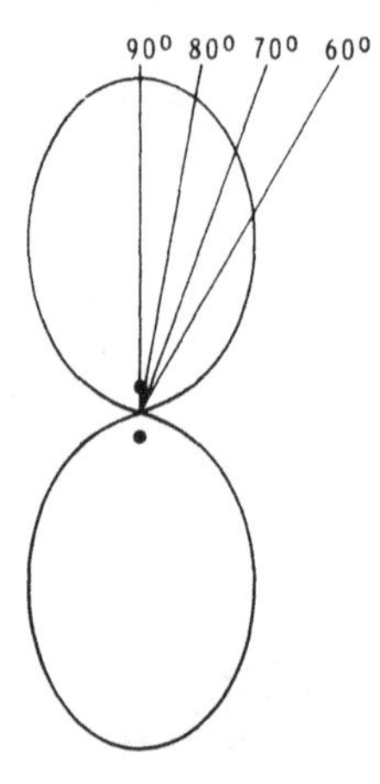

Fig. 5-4. Top view of the direction pattern of two vertical radiators fed 180° out of phase.

The method of powering the radiators is so simple that an antenna expert may doubt that it will work. Yet, careful checks with an SWR meter read less than 2 to 1 with very little change from 28.5 to 29 mc. Undoubtedly this could be reduced by carefully adjusting the length of the radiators. The radiators are spaced a bit more than a quarter wave in order to keep the radiation resistance as high as possible. Another advantage of the spacing is that the antenna can be used on 15-meters as well; making possible a two-band beam.

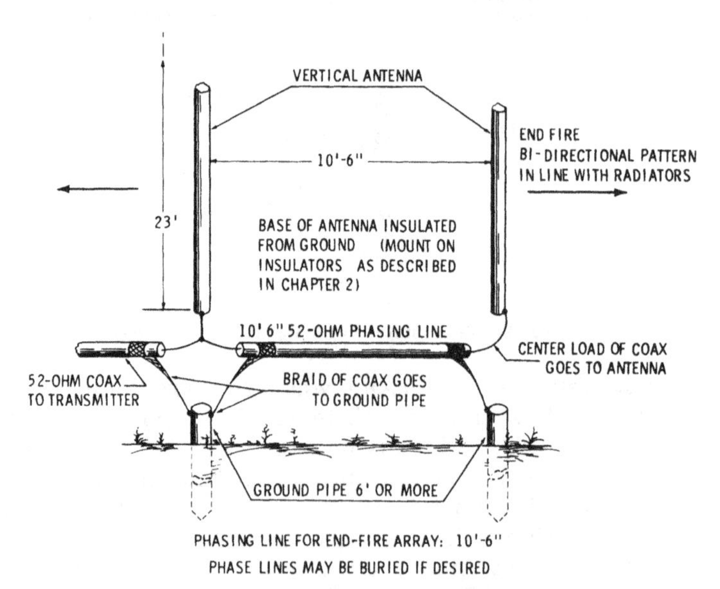

Fig. 5-5. The 50-Minute Special, 10-Meter Beam antenna.

The phasing line is an electrical half wave. (16 feet multiplied by .66—the velocity factor of standard, solid dielectric coaxial line. If you use the foam-type of coax, figure a .75 factor, or a 12-foot phasing section.) Putting the antenna on the air is simplicity itself. Cut the various parts to size, as shown in Fig. 5-5, mount the antenna, and hook it up to the transmitter.

Should any adjustment be indicated (for example, failure of the antenna to load-up sufficiently), the length of the coax feed line can be varied. Also, if an SWR meter is available and you are using commercial antennas as radiating elements, adjust the overall length of the radiators by sliding the telescoping elements up and down to achieve lowest SWR and maximum loading. The adjustment should be small—not more than a foot either way.

THE 15-10, 2-BAND BEAM

For the ham with modest power, the 15- and 10-meter bands have long been favorites, since even a 25-watt rig on these bands, plus a good low-angle antenna, will allow 1500-3000 mile contacts. In addition, overseas DX is good when conditions are right.

A 15-meter parasitic array with standard-size 22-foot elements becomes a fairly sizable antenna, and usually requires a commercial rotator. A vertical beam for the same band can be made up from two 23-foot verticals, as was described for the 50-Minute-Special. The radiators are fed (as was the 10-meter array) with a half-wave phasing section between them for an end-fire array.

The 10-foot 6-inch spacing between the two elements is approximately one-quarter wave at 21 mc. It is a bit more than needed on 28 mc, but although this somewhat reduces the gain, it is actually an advantage in that it makes the antenna broad-tuning—an important advantage on 28 mc (a band covering a large number of frequencies). It is also an ideal compromise dimension for using the antenna both as an end-fire and broadside array on 10 meters, as is covered later.

Fig. 5-6 shows the feed arrangement for 10 meters; and Fig. 5-7 shows how to feed the antenna on 21 mc. Notice that the difference lies in the length of the phasing line and that the feed method is slightly different on the two bands, 21 mc using a few turns of coil to improve the impedance match. On 10 meters, ordinarily, no tuning will be needed. The lower-frequency band, however, will require it.

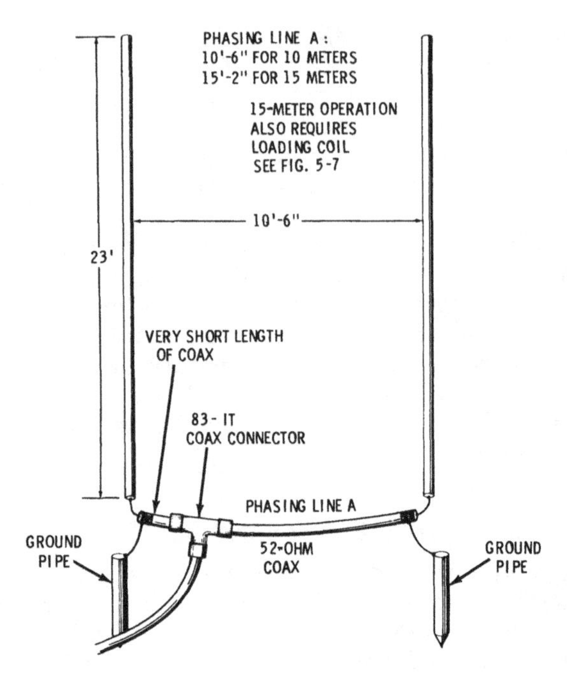

Fig. 5-6. The 15-10 Two-Band Beam antenna.

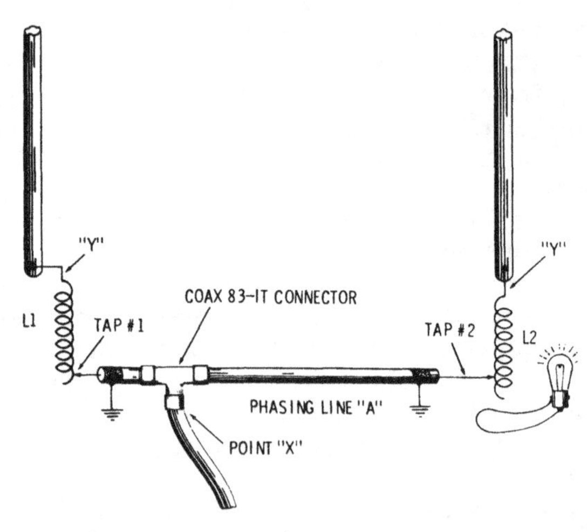

WITH 15'-2" PHASING LINE—ANTENNA
IS END-FIRE ARRAY ON 15 METERS

ON 10 METERS PLACE TAPS AT "Y" USE
10'-6" PHASING LINE FOR END-FIRE ARRAY

Fig. 5-7. Feeder hookup for 15 and 10 meters.

Tuning-Up on 15 Meters Without an SWR Meter

Here is a procedure, worked out after considerable experimenting, which makes possible tune-up without an SWR meter. The procedure is recommended even if you have a meter, since it insures that everything is within range for final adjustment with the SWR meter. Using the meter alone may lead you astray if things are hopelessly out of adjustment.

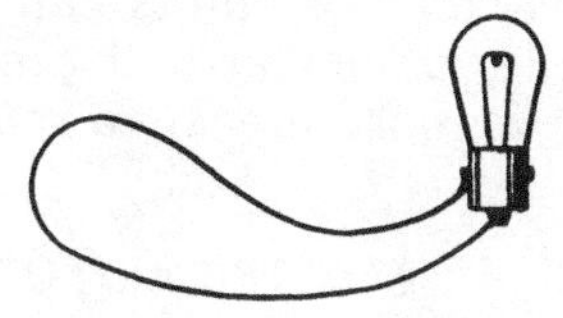

Fig. 5-8. Dial-lamp pickup loop.

1. Make up a pick-up loop with a No. 47 dial lamp as shown in Fig. 5-8.
2. With the antenna set up as shown in Fig. 5-7, disconnect phasing line *A* at the coax *T* connector. The purpose of this is to allow tuning up one antenna element first.
3. Turn on the transmitter, and adjust for modest output. Keep a helper at the transmitter to turn it on and off, and also to keep the transmitter on resonance.
4. Couple the pick-up look to loading coil L1. Adjusting tap #1, quickly determine the coil turn which gives maximum indication on the bulb.
5. Turn the transmitter off and place the tap on the point at which resonance was indicated.
6. Reconnect phasing line *A*, and place tap #2 on the same *number* of turns as tap #1.
7. Check the resonance of coil L2 as indicated by the bulb. If necessary, increase the transmitter output enough to get a good indication.
8. Now recheck the resonance of coil L1. It may be that there seems to be more or less output than on coil L2.
9. The final adjustment is to make small changes in both tap positions until the output on both coils appears to be the same. The adjustments are somewhat interrelated so that some experimenting is required.

On the test antenna for the hook-up shown, two turns were required on each coil.

Tuning With An SWR Meter

If you have an SWR meter available, tune the antenna as described. Then add the meter in the feed line at point *X* in

Fig. 5-7, or at the transmitter. Adjust the taps on both coils for minimum SWR.

10-Meter Operation

On this band no tune-up will ordinarily be needed. The only difference in the antenna lies in the length of the phasing lines. If the 10-meter phasing lines are fitted with proper plugs, the line can be quickly lengthened at any desired time. The 21-mc loading-coil tap points can be marked in some fashion (for example, by finger-nail polish on either side of the proper point) to make it easy to repeat the same adjustment.

ELECTRICAL ROTATING ON 10 METERS

The multiple-element vertical arrays discussed so far are all of the end-fire array variety, and give *figure-8* patterns in line with the direction of the two antennas. However, for 10-meter operation, the same antenna can be operated as a broadside array simply by lengthening the phase lines. Fig. 5-9 shows the two patterns available, and the length of the lines.

Such an antenna can be most useful on 10 meters, since, if oriented on the Europe-Australia axis, it will be effective for these areas during winter months. Then, as spring comes, the phasing lines can be lengthened and pattern *B* made available for working South America and South Africa, which usually come in better as the season progresses into summer.

A UNIDIRECTIONAL ANTENNA FOR 21 MC

Generally, on 10 meters, a bidirectional antenna is perfectly satisfactory, since, due to the skip effect, strong signals are seldom received from opposite directions at the same time. On 21 mc, however, this is not necessarily true. There is a real advantage to an antenna which transmits and receives essentially one-way, thus yielding the cardîoid pattern of Fig. 5-10. Also, the gain is stepped up to approximately 5.5 db. With the antenna spacing shown in Fig. 5-11, this can be achieved simply by feeding one radiator 90° out of phase with the other. Changing direction simply involves going outside and swapping ends with the phasing line.

For a more deluxe installation, feed lines of equal length can be brought inside the house and switching accomplished at that point by swapping cable ends, as shown in Fig. 5-12 (or by a switching arrangement as shown in Fig. 5-13). This arrangement yields a bonus in that it makes it easy to feed the two

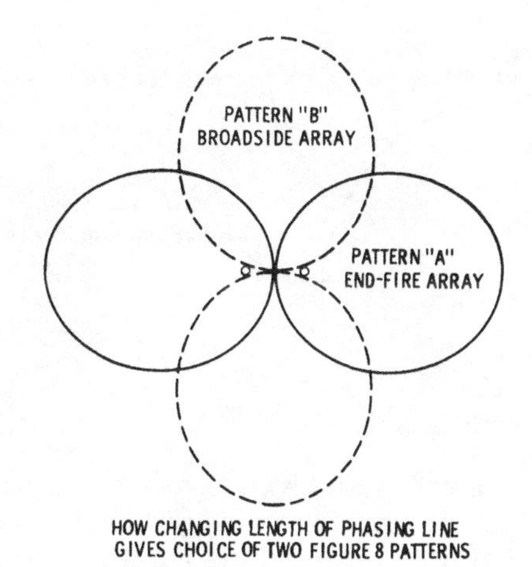

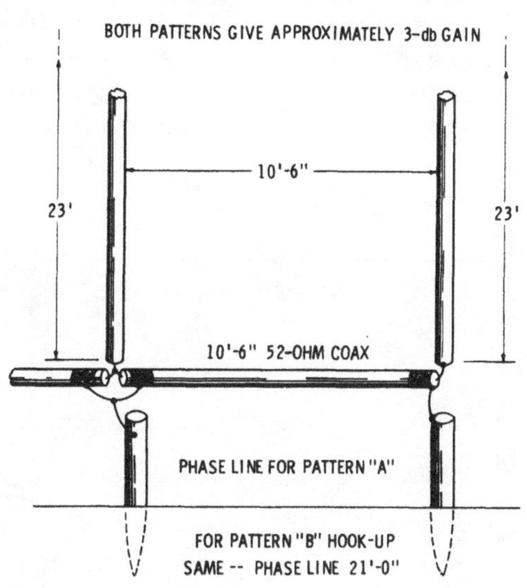

Fig. 5-9. Electrical rotating of the 10-Meter Vertical Beam antenna.

radiators in phase, thus achieving a broadside array as was done with the 50-Minute-Special, 10-meter beam described earlier. Simply omit the 52-ohm quarter-wave phasing line, thus insuring that the two antennas are fed in phase.

See Fig. 5-14 for a switching arrangement to give you the same choice of two unidirectional cardoid patterns and one *figure-8* bidirectional pattern.

The nice thing about this system is that rotation can be accomplished in your operating room simply by switching around a few cables. And if the uni-directional patterns are oriented

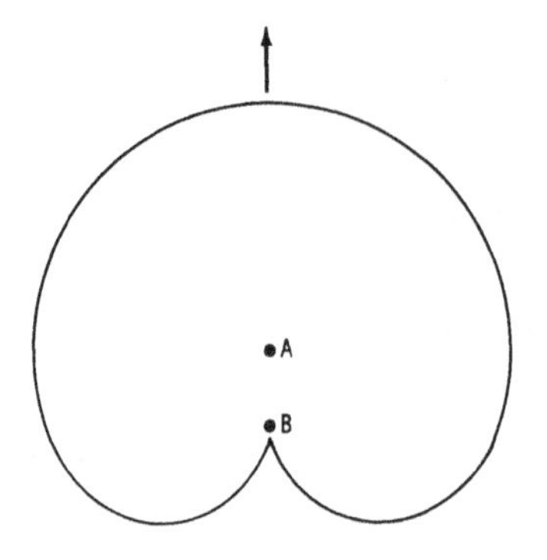

Fig. 5-10. Top view of radiators *A* and *B*. Cardiod pattern for both when fed 90° out of phase.

on the Europe-New Zealand axis, with the broadside pattern on the Tokyo-Rio de Janeiro axis, good world-wide coverage will be obtained simply by changing cable connections.

The same type of antenna can be used on 20 meters, too. Fig. 5-15 shows the dimension of the feed lines and the overall layout as well. Obtaining a low SWR reading on 20 meters may be difficult with only the loading coil shown in Fig. 5-16. If so,

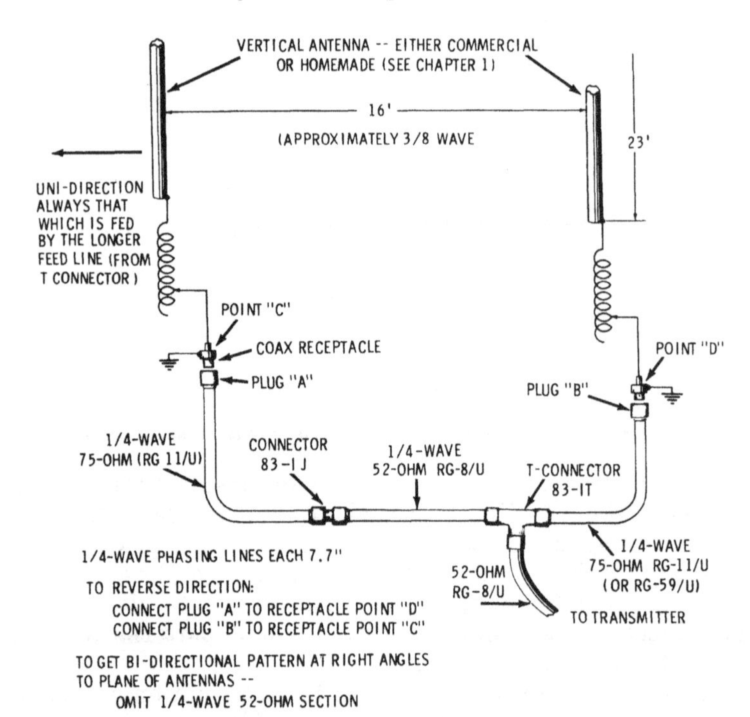

Fig. 5-11. A 15-meter array giving a choice of two unidirectional patterns (5-db gain) plus a Figure-8 bidirectional pattern with approximately 3-db gain.

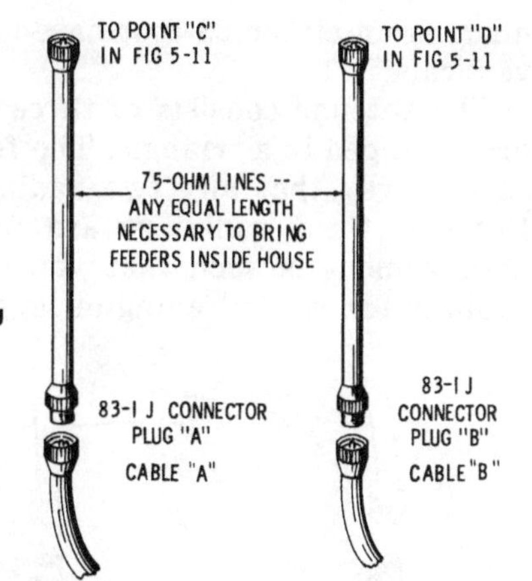

Fig. 5-12. Extending the feed lines to bring the switch point inside the house.

use the hookup in Fig. 5-15, which is the same one described in connection with the 80-meter vertical in Chapter 2. Tune-up is the same. Adjust one tap to vary the electrical length of the coil until resonance is achieved, then adjust the tap carrying the feeder to achieve an impedance match.

DELUXE 40-METER BEAM

Fig. 5-17 shows a 40-meter beam which will give 4 db or more in a choice of six different directions—and the directions can be chosen simply by turning a switch. The beam can be

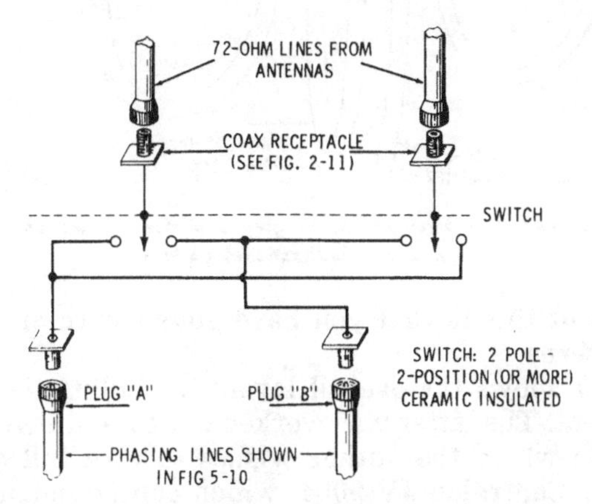

Fig. 5-13. Switching arrangement to accomplish direction change.

made from either 23-foot base-loaded verticals or full 33-foot verticals.

The antenna consists of three quarter-wave vertical radiators arranged in a triangle. The feed system can be switched in such a way that any two radiators are in use at one time. Further, these radiators are fed 90° out-of-phase, and the arrangement is such that you can choose which of the two radiators in use is leading or lagging the other by 90°. The up-

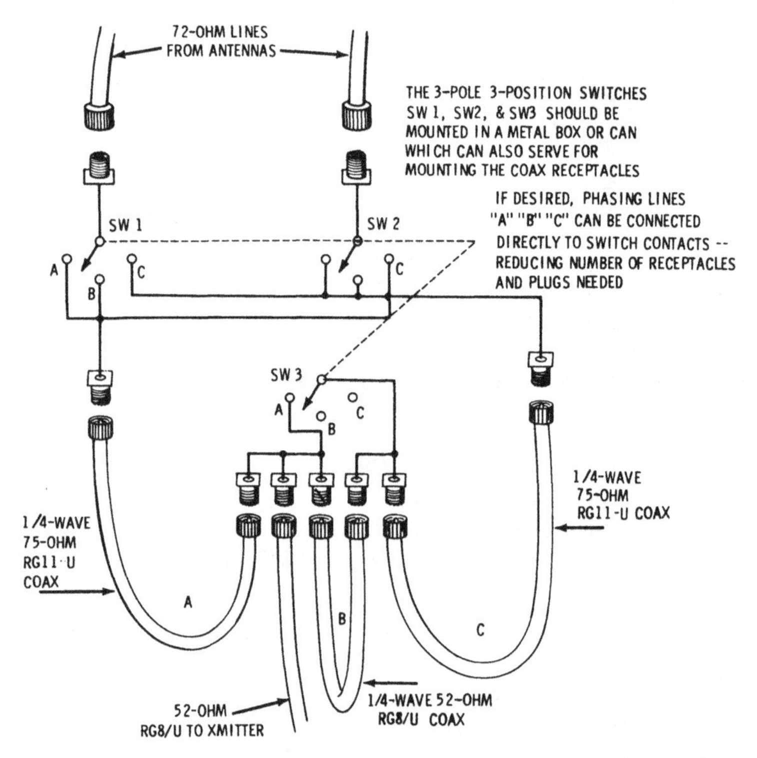

Fig. 5-14. Switching arrangement to give a choice of two unidirectional and one bidirectional pattern.

shot of all of this is that you have your choice of 6 unidirectional patterns.

Fig. 5-17 shows the overall layout, including the switching arrangement. The latter was worked out by a Midwestern ham, W9OKN, to whom the author wishes to give full credit. The switch is a Centralab JV-9004, which can be mounted in any small metal box, which also serves to mount the coaxial receptacles.

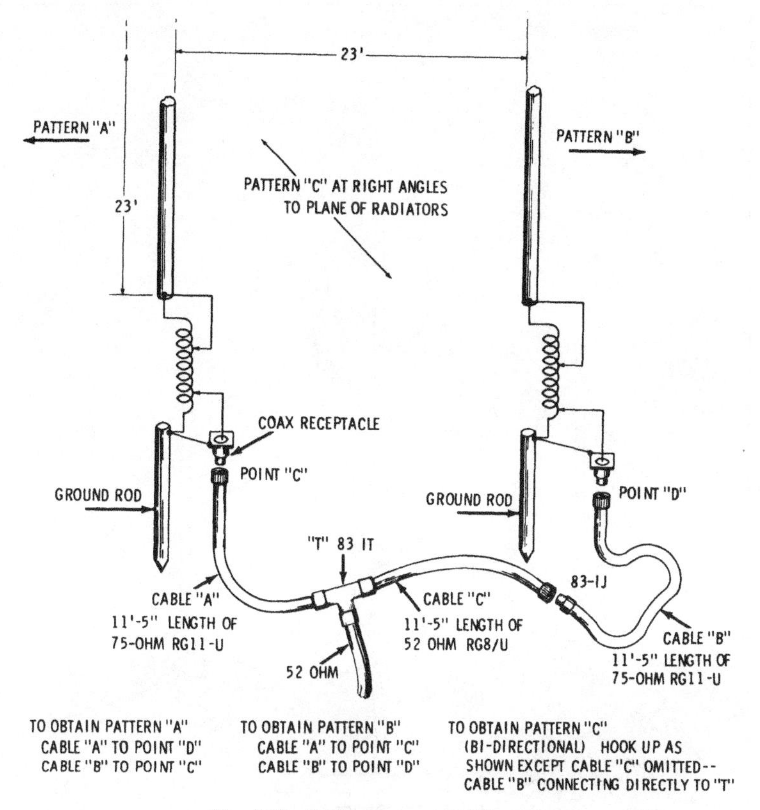

Fig. 5-15. A 20-meter vertical beam.

Tuning The Antenna

Assuming you are using 23-foot verticals with a loading coil at the bottom, adjustment is fairly simple. Hook your SWR meter in series with the coax feeding one of the verticals, then

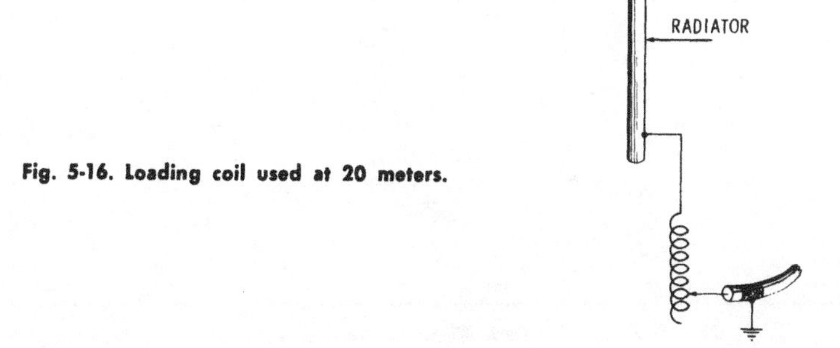

Fig. 5-16. Loading coil used at 20 meters.

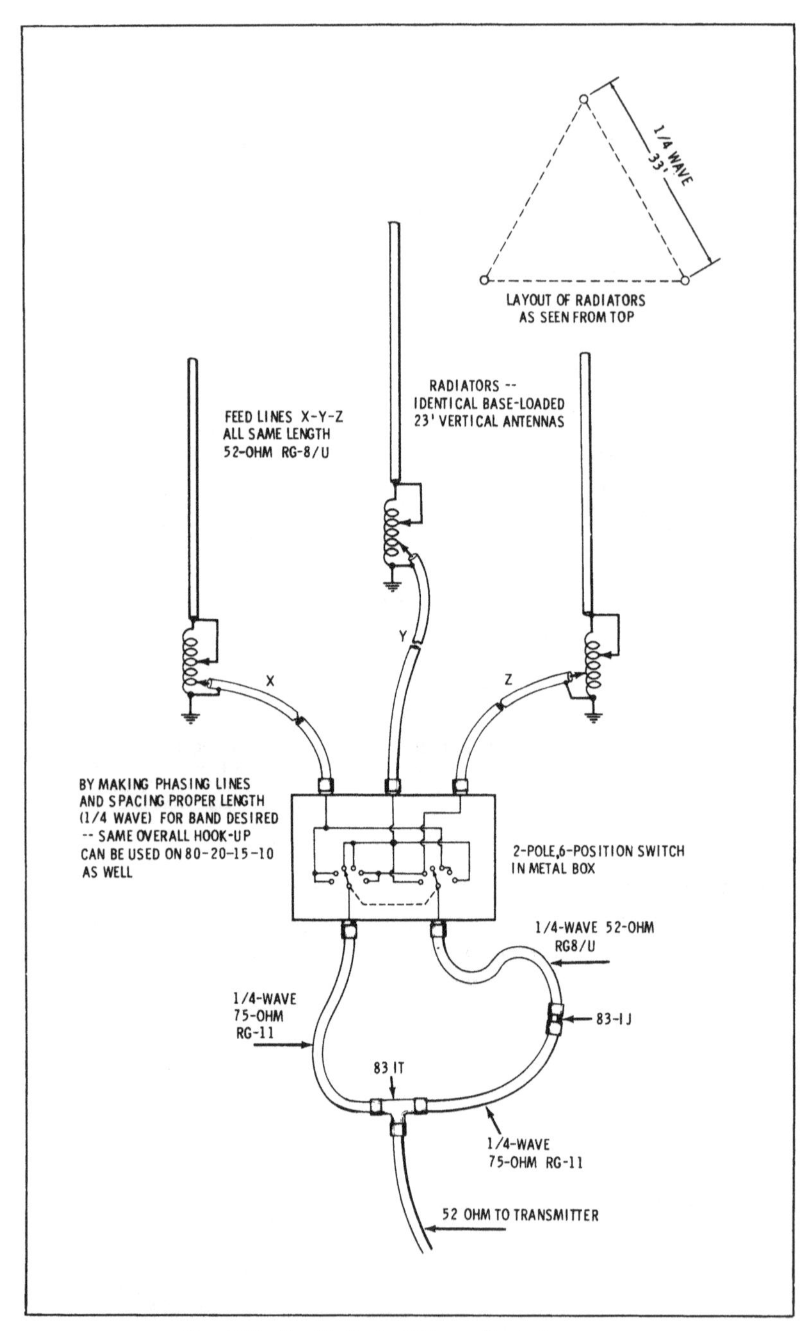

Fig. 5-17. The Six-Pack 40-Meter Vertical antenna.

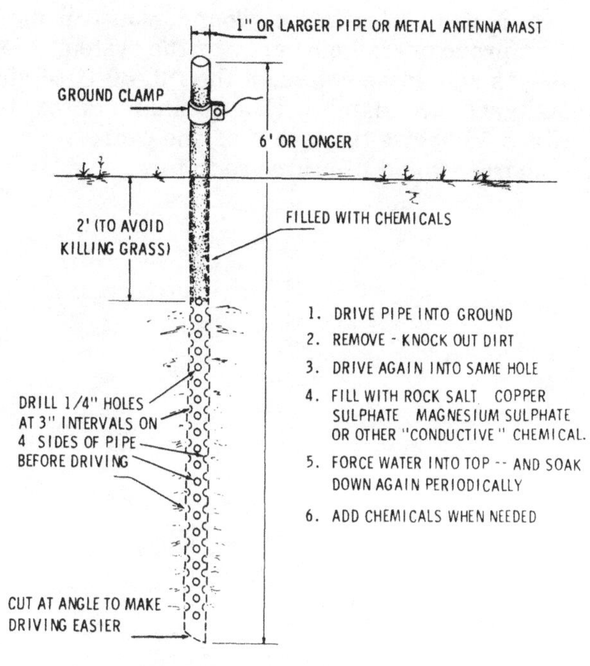

Fig. 5-18. A chemical ground.

follow the same procedure outlined in Chapter 2 for tuning up the 80-meter vertical radiator. Resonate the overall system (either with a grid-dip oscillator, or by supplying power by tapping up a few turns from the grounded end) and then tap the coax line on at the point which gives lowest SWR. The adjustments are interrelated—so some experimenting is necessary.

After each vertical radiator has been tuned up, feed the radiators two at a time. SWR may have changed somewhat. If so, make some adjustments as are necessary to bring them in line.

Grounding The Antenna

It does not make any sense to build a vertical beam of this type in order to achieve a power gain only to throw it away in a poor ground; this is easy to do. A high-resistance ground can soak up all the power gain; so don't even consider this type of antenna unless you are willing to go to the trouble of building a good ground. Another chemical ground is shown in Fig. 5-18. It is quite effective, but even so, should be backed up with some radials if possible.

Of course, instead of using a ground, you can use radials, which will improve overall efficiency of the system. Ideally, the radiator length should be raised to the full 30 ft to aid in getting a good impedance match with a minimum of loss in a loading coil. Fig. 5-19 shows the layout of one radiator; a complete system, of course, would use three radiators.

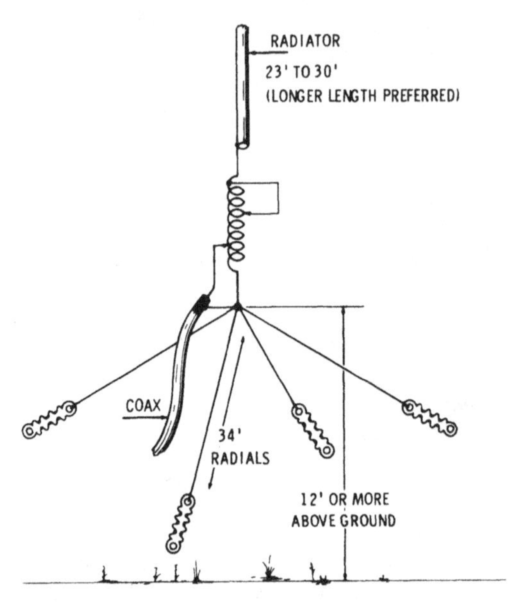

Fig. 5-19. Ground-plane element for the 40-Meter Beam.

80-METER VERTICAL BEAM

With everything scaled up to the proper dimensions, an 80-meter vertical beam is possible too, of course. To build one, get as much radiator up as possible and use a loading coil at the base plus a good ground system. Use the same overall layout as shown in Fig. 5-11 to allow a choice of two unidirectional patterns (achieved simply by swapping ends with cables). An even better arrangement is that of using two 80-meter ground planes, spaced one-fourth wave (66 feet) apart. The ground planes can then be fed with the same hook-up as shown in Fig. 5-19.

For 80 meters, the dimensions are: radiator—up to 55 feet; radials (if used)—66 feet; one-fourth wave matching sections—246 times the velocity factor of the line divided by the frequency in megacycles. Standard coax has a velocity factor of .66, "Foam" type—.75.

CHAPTER 6

Compact and Indoor Transmitting Antennas

Most hams dream of an antenna farm with acres and acres of open space, except for 100-foot tall pine trees to use as antenna masts. Unfortunately, such dreams are seldom realities, and many a ham has to cope with such practical problems as trying to get on the air despite a landlord who takes a dim view of a beam on the chimney. It is for such luckless souls that this chapter is written. The antennas described are all either miniature in size, or are of a type which lend themselves to use indoors.

WINDOW-SILL RADIATOR

Let's start with an antenna which is useful for the worst possible situation—the ham living in a college dorm or in an

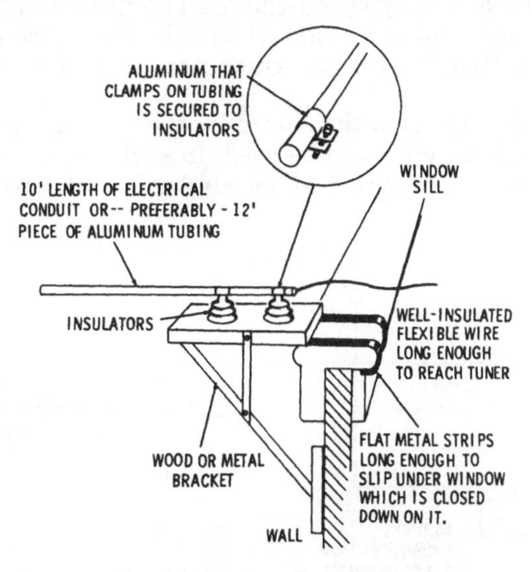

Fig. 6-1. Window-sill antenna.

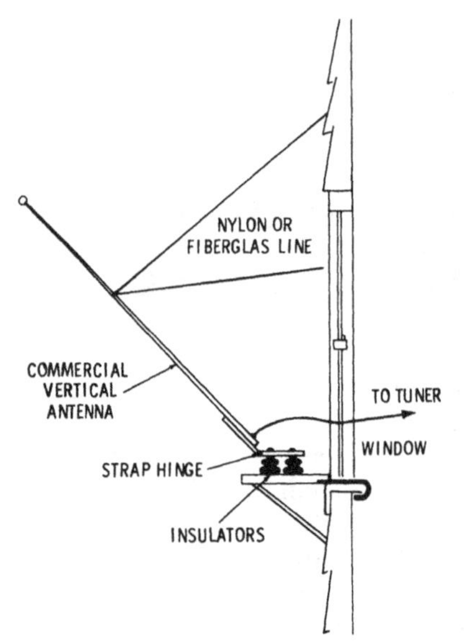

Fig. 6-2. Another method of constructing a window-sill antenna.

apartment house where there is no chance to string a wire. One solution in such a situation is to extend a rod out the window of a room, and then base-load the rod with a coil of sufficient inductance so that the rod will resonate on the desired band. Fig. 6-1 shows the rod hook-up. Exact mechanical arrangement will depend on the type of window. The important thing, however, is to work out a method which will mount the rod securely enough so that it will not come crashing down in the first breeze.

The longer the rod, the better it is; a longer, manufactured, vertical-type antenna can be used. In such a case, however, the antenna is best mounted at an angle and supported by some

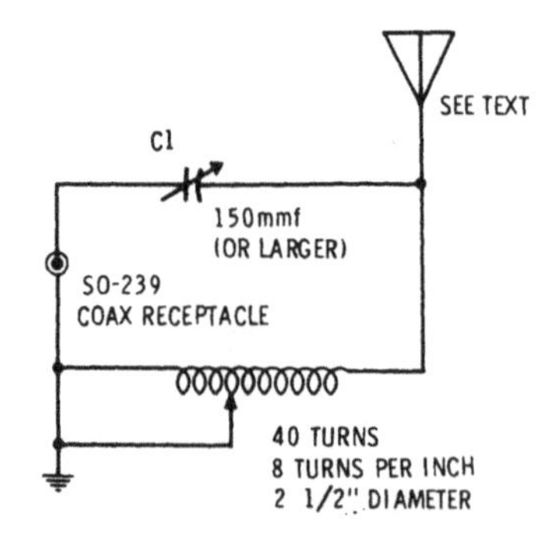

Fig. 6-3. Window-sill antenna tuner schematic.

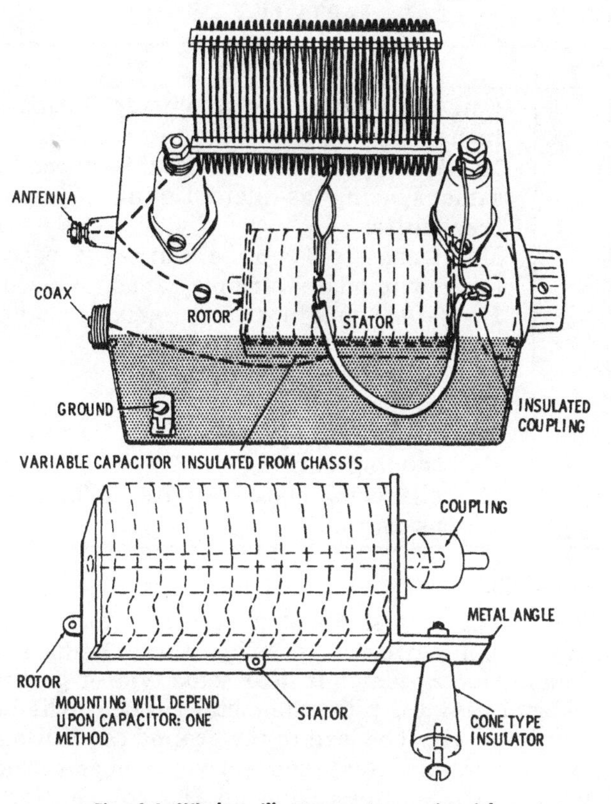

Fig. 6-4. Window-sill antenna tuner pictorial.

insulating line, or, small guy wires and insulation as shown in Fig. 6-2.

To tune up such antennas requires an impedance-matching tuner such as the one shown in Figs. 6-3, 6-4, and 6-5.

Fig. 6-5. Window-sill antenna tuner.

PARTS LIST

Quantity	Description
1	Coil (40 turns 2½" diameter 8 turns per inch).
1	150-mmf (or larger) variable capacitor of same spacing as plate tuning capacitor in transmitter.
1	Metal chassis 3 × 6 × 4 (larger if necessary to accommodate variable capacitor used).
1	Insulated coupling for variable capacitor shaft.
1	SO-239 coaxial receptacle.
2	Stand-off insulators.
2	Feed-through insulators.
1	Coil-tapping clip.
	Miscellaneous hook-up wire, 6-32 machine screws and nuts.

Tune-Up Procedure

An SWR meter is just about a necessity for this type of antenna system. Likewise, if possible, the antenna should be mounted on a window which is near some type of ground. This can be a steam radiator pipe, the metal frame of the building, a cold water pipe, etc. The lead to the ground connection should be as short as possible. Next, connect your tuner to the transmitter with a coaxial line. Tune up the transmitter on the desired frequency in the desired band. Set the SWR meter to read reflected power. Rotate C1 in Fig. 6-3 to see if you get a dip in the meter reading. If not, tap the coil a turn at a time until you get a dip. Then adjust the coil a fraction of a turn at a time until you get a *pronounced* dip, ideally so deep a dip that the meter indicates very low SWR. The combination of C1 setting and coil tap which gives the lowest SWR is the proper tune-up. On low frequencies (e.g., 80 meters) tuning is very critical—becoming more broad on the higher frequencies.

The efficiency of the window-sill radiator can be improved considerably by a technique known as top-loading-with-a-capacitive-hat. One simple way to do this is to use one-half of a conical TV antenna on the outer end of the rod. How much of the conical antenna to use will depend upon how husky a radiating rod you have. If you are using a 10′ aluminum TV mast as a radiator, you can use the whole thing; otherwise, better cut it down so that it is perhaps 2′ across. Modifying the elements by screwing the scraps together to form a kind of lopsided wheel (Fig. 6-6) is also a good idea.

The usefulness of the window-sill antenna can be improved by mounting the rod on a strong hinge mounted on a single strong insulator, and then using nylon or *fiberglas* guys to support it and to allow shifting direction by pivoting the hinge on its mounting insulator.

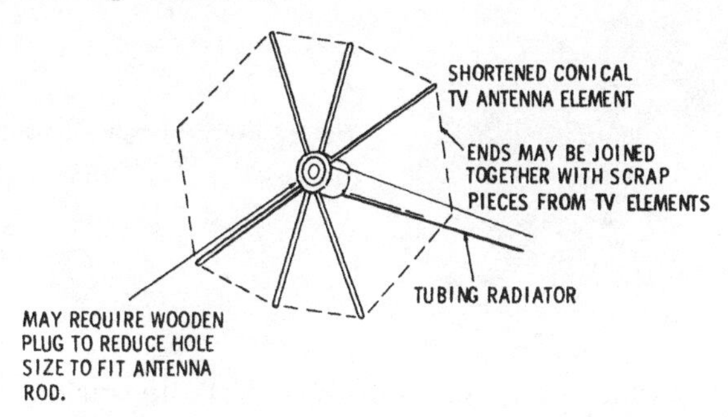

Fig. 6-6. Top-loading radiator for improved efficiency.

USING A RANDOM-LENGTH WIRE

A very similar tuner to that used for the window-sill antenna can also be used to put power into a random length of wire. So, if you have the space to do it, get out as long a wire as you can. The tuner for this purpose uses the same parts as that for the window-sill antenna tuner. However, the circuit is rearranged slightly. Also, provision is made for hooking the variable capacitor to either end of the coil (Figs. 6-7 and 6-8).

For most lengths of wire, clip *A* from C1 goes to point *C* on the coil. However, if you cannot get the antenna to tune up properly, try moving the clip to point *D*.

Tune-Up Procedure Using Bulbs

The best way to tune up a random length of wire is to follow the procedure using an SWR meter as was outlined in the case

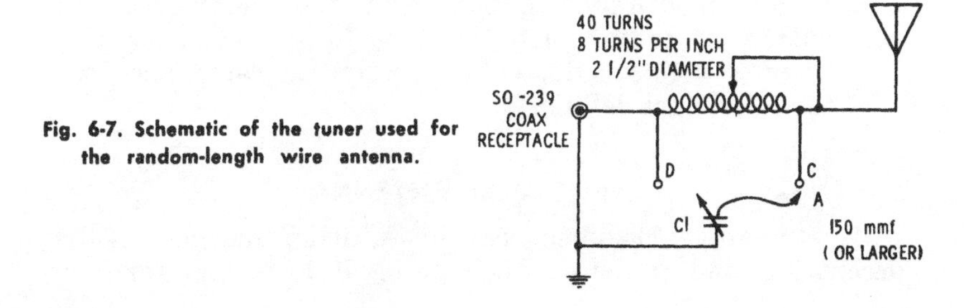

Fig. 6-7. Schematic of the tuner used for the random-length wire antenna.

of the window-sill tuner. However, if you are working with a wire 20 feet or more in length, the tune-up is not quite as critical as with a shorter rod. For this reason low-cost bulbs can be used to indicate resonance.

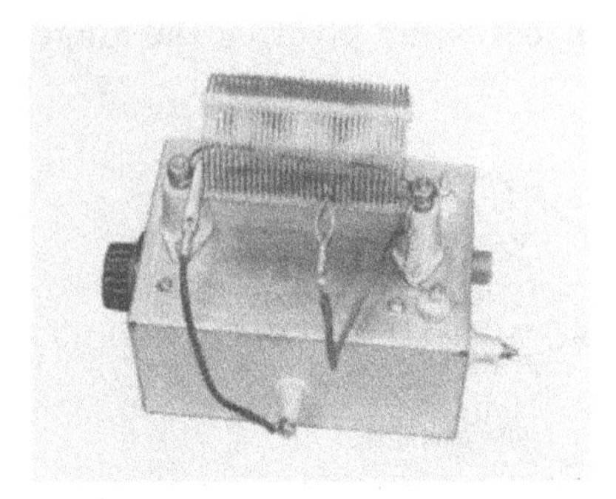

Fig. 6-8. Random-length wire antenna tuner.

1. Hook a dial light or flash-light bulb in series with the antenna to be used. As was explained in Chapter 3, in case the power of the transmitter requires it, use a Christmas-tree light bulb, or parallel several flashlight bulbs.
2. Connect the transmitter to the tuner and tune the transmitter to the desired frequency. Be sure to note the input of the transmitter as indicated by plate current.
3. Applying the coil-tap clip to the coil; by experiment determine which combination of capacitor setting and number of coil turns gives the most output as indicated by the brightness of the bulb (or RF meter if you have one). It is very important, of course, that the input to the transmitter be held constant.
4. As mentioned previously, you may encounter a length of wire which is of an impedance that can be tuned better with the alternate arrangement provided. Also, in some cases, you may have an antenna length which puts the feed point close to a voltage maximum. In this case, a neon bulb with its leads soldered to the antenna post, and with a grounding plate provided to give capacity coupling to the glass bulb, will be a better indicator. See Chapter 3 for more details.

ATTIC WIRE ANTENNA

By stringing a random wire in an attic, you get a fairly decent all-band radiator. Zig-zagging it back and forth be-

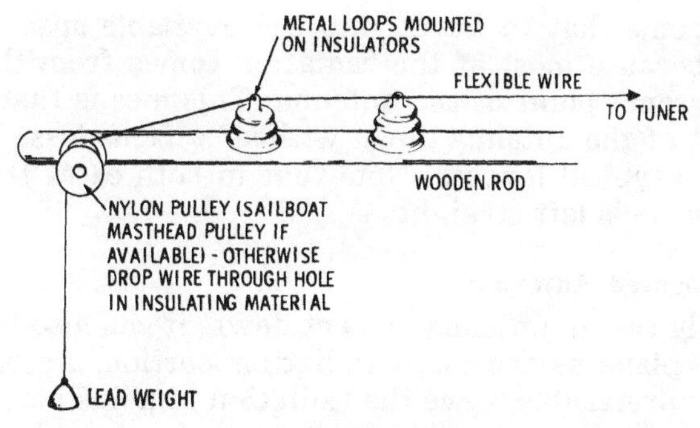

Fig. 6-9. Fish-line antenna.

tween roof rafters is one practical way to get out quite a bit of wire in a fairly limited space. On the 80- and 40-meter bands, getting out a lot of wire is an advantage.

FISH-LINE ANTENNA

Another way to design a random-wire, out-of-the-window antenna, is shown in Fig. 6-9. The antenna, of flexible wire and fitted with a weight, is simply dropped out of the window, held away from the building by a short wooden pole. Be careful in using such an antenna that the lead weight doesn't swing into a window below.

ADDITIONAL INDOOR ANTENNAS

Antennas like the coax-fed dipole in Chapter 1 perform well indoors too, and such an antenna can often be mounted in an attic for 20, 15 or 10 meters. If you have to, you can bend the

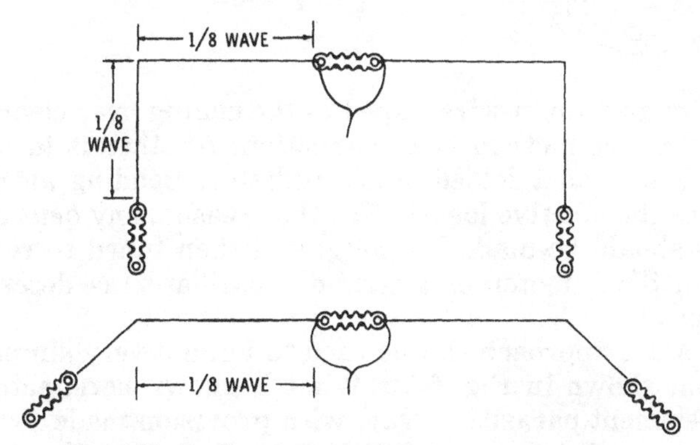

Fig. 6-10. Bending antenna elements to save space.

antenna somewhat to get it into the available space. This is possible because most of the radiation comes from the center (high-current) point of the antenna. This means that you can drop part of the antenna down without serious loss. Fig. 6-10 shows two typical layouts. Note that in both cases the middle quarter wave is left straight.

Ceiling-Mounted Antennas

Not only can an antenna be bent down, it can also be bent in the same plane as the main radiating portion. Actually, this method is preferable, since the radiation remains horizontal or vertical, depending, of course, on how the antenna is mounted. As a matter of fact, the whole antenna can be bent in a kind of open loop as shown in Fig. 6-11. Such an antenna is directional toward the feed point; so, if mounted on the ceiling of a room, orient it in the favored transmitting direction.

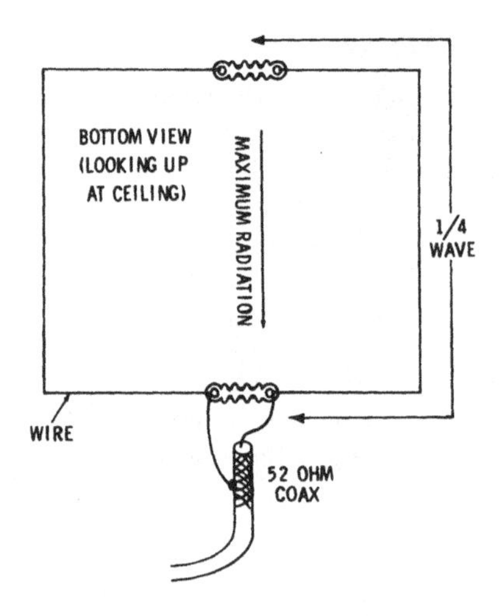

Fig. 6-11. Ceiling-mounted loop antenna.

Heavy aluminum wire, taped to the ceiling over clear plastic spacers is one method of construction. Another is to use 300-ohm TV lead as a folded-dipole radiator. Bending an antenna shortens its effective length. For that reason, any *bent* antenna ideally should be made 5% long, and then tuned to resonance with an SWR meter or a grid-dip oscillator as described in Chapter 3.

The same approach can be used to build a very simple beam like that shown in Fig. 6-12. What is shown here, actually, is a two-element parasitic array, with provision made for tuning up the parasitic element as either a reflector or director. This gives a practical way to reverse antenna direction and the whole job can be done simply by throwing a switch. Here is the tune-up procedure:

1. With switch S1 open, and by using capacitor C1, tune in a signal in the direction of the parasitic element. (For example, if the parasitic element is on the west side of the room and the radiator is on the east, tune in a station from the west.) Tune for maximum *S*-meter reading on your receiver.

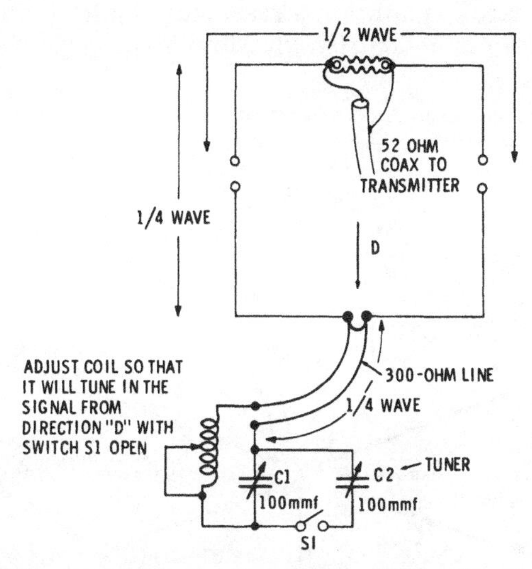

Fig. 6-12. Ceiling-mounted beam with reversible direction (for 10-15-20 meters).

2. Close switch S1, and tune in a station from the opposite direction with C2.
3. Now, simply by opening and closing switch S1, you can reverse the direction of the antenna.

MINIATURE PARASITIC BEAMS FOR 21 and 28 MC

Most of the vertical beams already covered do not give full 360° coverage. As explained, this is not a serious disadvantage, since seldom is 360° coverage needed (even for working DX over a wide portion of the world). However, a mechanically-rotated beam does enable you to zero-in precisely on the desired station. Also, in certain types of locations (for example, where buildings are close together), mounting a vertical antenna at ground level is not very efficient. Full-size commercial three-element beams are a good answer to such problems. However, the bulk and cost of such beams are negatives for many hams. For them, a low cost, miniature beam may be the answer.

TV-SPECIAL MINIATURE BEAM

Combining several elements in a parasitic array tends to lower the radiation resistance and make the beam more critical in tuning. When the size of a beam is reduced, all of these problems become more critical. A miniature beam must be designed with special care to maintain a reasonable radiation resistance. However, there is a surprisingly simple and practical answer.

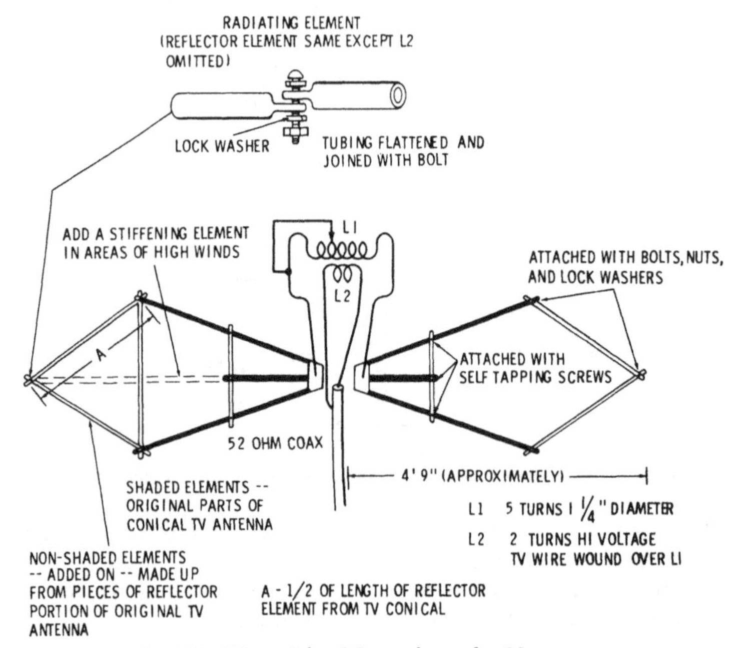

Fig. 6-13. TV special, miniature beam for 10 meters.

A ham, K6OFM, several years ago developed a type of miniature dipole which he called the *Wonderbar*—actually a modification of the fan radiator which has been used on VHF for many years. Subsequently, other hams, including W2AWH, worked out simple 21- and 28-mc beams using the Wonderbar principle. Fortunately for the ham with a thin budget, it is possible to build antennas of this type by modifying standard conical TV antennas. In fact, two conicals provide everything needed to make a 28-mc beam except for some small coils and odds and ends of hardware. Your bill for the whole thing will generally be less than $10.00. Three conicals plus a standard aluminum TV mast will furnish the parts for a 21-mc miniature beam.

Figs. 6-13, 6-14, and 6-15 show the layout for a 28-mc beam. In the actual assembly of the fan-shaped radiating element, the original antenna portion (that which is attached to the plastic insulator) is enlarged by adding sections to it which are removed from the part of the conical which was originally the reflector section.

Fig. 6-14. TV special mounted.

The exact dimensions of the new element are not particularly critical, which is fortunate because the TV antennas vary in dimension. In choosing an antenna for the modification, buy the best quality you can find. The best generally will cost only a dollar or so more than cheaper ones, but they are usually larger, have tougher tubing, and better hardware.

Note in Fig. 6-13 that one of the reflectors is cut in two to provide the kite-shaped end for the radiator, and the second reflector provides some needed cross bracing at two points. If you follow the construction method shown, you will use up all but a couple of inches of the original antenna elements in making one of the fan-shaped dipole radiators. As shown, the various pieces of tubing are assembled either with self-tapping screws (in the case of the braces) or with nuts, bolts, and lock washers. Ideally, the latter should all be of aluminum; however, cadmium-plated steel parts can be used instead.

In taking the original TV antenna apart you may find that some of the elements are riveted to the boom. The easiest way to remove the rivets is to drill them out.

Also, the job of bolting together the various parts which make up one of the dipole radiators is greatly simplified if you can lay the radiator flat on the floor of your basement or garage. This may require removing a bolt or rivet which originally secured the whole antenna assembly to the boom.

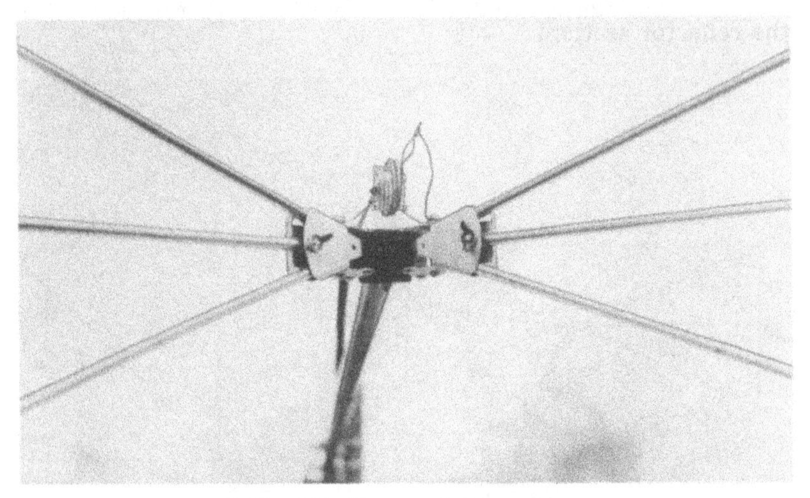

Fig. 6-15. Close-up of TV special.

Assuming you are using identical TV antennas as your source for parts, make up two identical fan-shaped dipoles. Drill new holes in the boom of each so that you can mount the *U* mounting clamp at the end of the boom. The idea is to stack one of the booms above the other as shown in Fig. 6-16. Be certain to leave enough overlap so that you can put on a *U* clamp to obtain a strong mechanical bond between the two booms.

Resonating the Elements

As you may have suspected, the overall length of the radiators is considerably shorter than that which would ordinarily be expected for a half-wave dipole on 28 mc. The fan-shaped radiators help the cause since they are resonant at a lower frequency than would be true of a single rod. Even so, we have to piece out the radiator in order to get the desired length.

This is done by adding a loading coil between the two segments of the antenna as shown in Fig. 6-13. This coil also provides a simple way of applying power to the radiator from

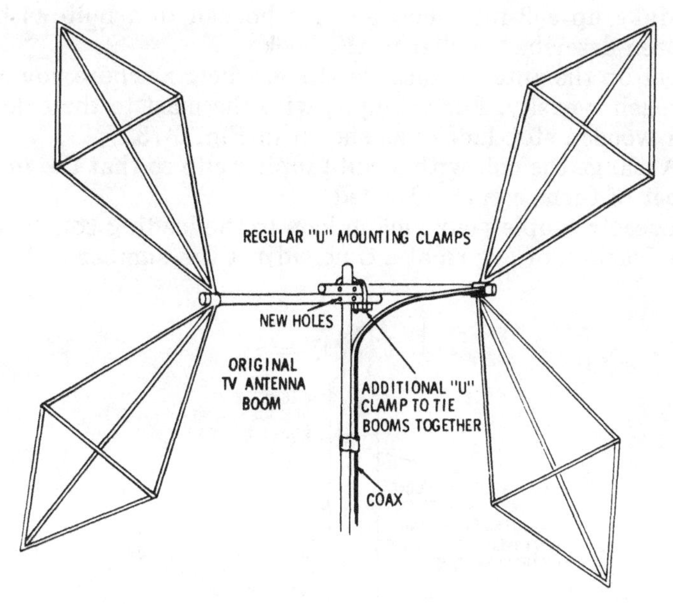

Fig. 6-16. Mounting detail of TV special, miniature-beam antenna.

standard 52-ohm coax. The reflector is identical to the radiator except that it has no coil for applying power—the reflector is simply a parasitic element.

For both the radiator and reflector, the loading coil can be five turns of #14 bare wire, wound so that it is 1¼″ inside diameter. The coil should have each wire spaced the width of the wire (Fig. 6-17) and the coil should be strengthened with strips of household cement (as described in Chapter 1 in connection with building the coil for the *Novice*-90 antenna). Even better, particularly for a rig of over 100 watts, would be a self-supporting coil of heavy wire or copper tubing.

The coil is connected between the two segments of the antenna elements. Over the center of the coil, wind a two-turn winding of insulated high-voltage wire of the type used to carry the high voltage leads to TV picture tubes. This provides a coupling coil for hooking up the feeder line. The antenna should then be assembled as shown in Figs. 6-14 and 6-16.

Tune-Up Procedure Without Instruments

Any of several different instruments can be used to tune up the TV-Special miniature beam. However, an acceptable tune-up job can be done with nothing more elaborate than the dial-light bulb and loop.

1. Make up a 2-inch loop of wire hooked to a bulb as has been described earlier in the book.
2. Set up the antenna outside at some height where you can reach it easily. For example, wire the mast to the side of a wooden step-ladder as shown in Fig. 6-18.
3. Arrange the coil with a coil-tapping clip so that the number of turns can be adjusted.
4. Loosely couple your pickup loop to the loading coil. Then, a fraction of a turn at a time, adjust the number of turns

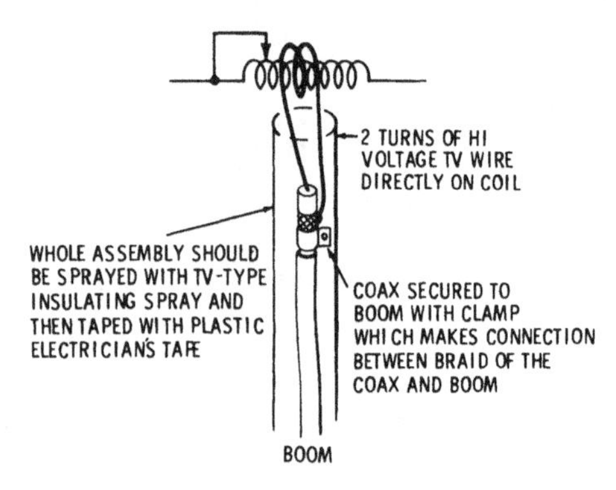

(A) Pictorial.

(B) In mounted position.

Fig. 6-17. Feed line and coil detail.

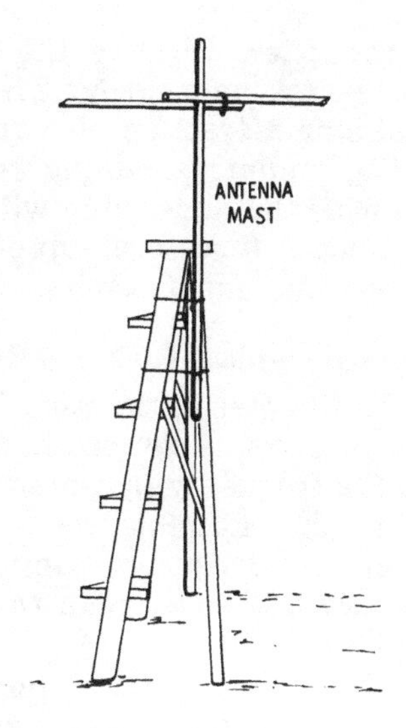

Fig. 6-18. Using a stepladder to facilitate antenna tuning.

so that the bulb gives the brightest glow. This indicates that the radiator is tuned to resonance.

5. Once the proper point is found, remove the tapping clip and solder the shorting wire directly onto the coil at the tap point.
6. A fairly good job can be done tuning up the reflector simply by making its loading coil two turns larger than the radiator. However, you can get a bit better tune-up by making a half-wave dipole out of a scrap of 300-ohm TV feed line with a bulb hooked in the center (Fig. 6-19). Mount this pick-up antenna out in front of the radiator of the miniature beam and then adjust the reflector for maximum brilliance of the bulb on the pick-up antenna.

Tune-Up Procedure with an FSM

A field-strength meter is one practical way to tune up the miniature beam. First, arrange it so the antenna can be fed power from a transmitter. Mount the antenna on a step-ladder, or some other type of wooden support, and power it with the

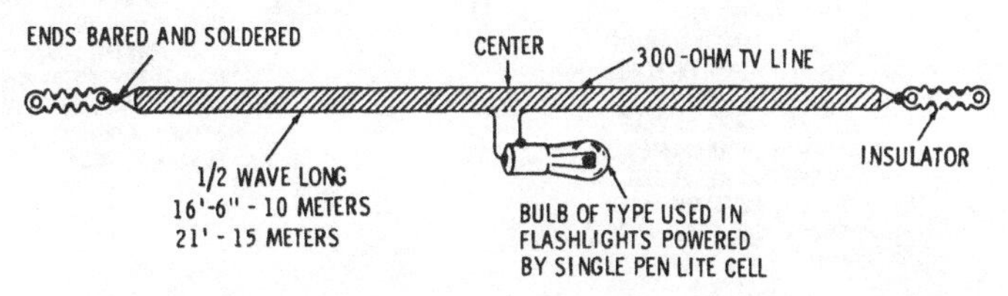

Fig. 6-19. Pickup antenna for adjusting reflector length of beam.

transmitter. Observe the output reading on a field-strength meter set up nearby. Then, adjust the loadnig coil on the radiator a fraction of a turn at a time until, with the transmitter input remaining constant, you get maximum output. Now do the same thing with the reflector tuning for maximum output. A final touch-up of both coils will probably be needed to get maximum results.

Tune-Up with a GDO and SWR Meter

In Chapter 3, the use of a GDO to get antenna elements on resonance was covered in detail. In this case, turn the radiator to the frequency in center of the most desired 200 kc of operation. (For example, set to 28,700 if 28,600 to 28,800 kc is the preferred frequency range.) Then peak the reflector at a frequency 5% lower than the desired frequency. In the example of 28,700, 5% is:

28700 kc	28700	27,265 kc—reflector frequency
.05	1435	
1435.00 kc	27265	

Tuning up in the preceding manner, if done accurately, should be close to ideal. However, SWR can usually be improved by adjusting the spacing (and perhaps adding or subtracting a turn) of pick-up coil L2. The object, of course, is to achieve the lowest possible SWR. You can also check resonance

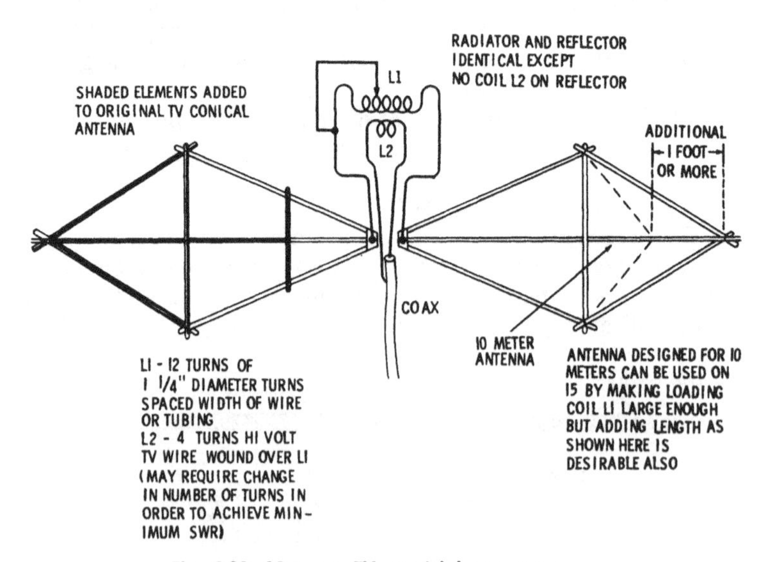

Fig. 6-20. 15-meter, TV special beam antenna.

of the radiator by adjusting the center loading coil for minimum SWR.

Tune-up on 21 mc follows the same procedure as 28 mc; the loading coil is simply larger. Ideally, the elements should also be enlarged, (See Fig. 6-20), and also the boom should be lengthened. A 10-foot aluminum TV mast makes an ideal boom for the 21-mc miniature beam. Short lengths of the original boom can be bolted on top of the TV mast boom.

What about frequencies other than 21 and 28 mc? The lower frequencies require elements too large to make from TV antenna parts. However, if you wish to experiment, here is a Wonderbar type of antenna with which one has reported success. Fig. 6-21 shows a dipole. Undoubtedly, however, two elements could be combined to form a beam exactly as was done with the TV-Special beam. A 10-foot boom would be minimum, and a 16-foot boom would be desirable.

All of the miniature beams described above can be rotated very easily with a TV rotator. Or, you may want to build a simple rotator yourself.

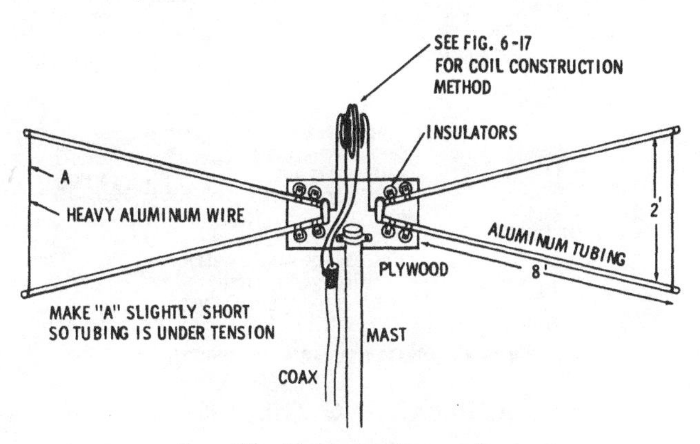

Fig. 6-21. Miniature 20-meter dipole.

HOMEMADE ROTATOR

This rotator is entirely mechanical, yet in practice it has proven highly useful. You can build it for as little as $6.00, if you have any luck finding parts bargains.

The rotator operates by means of a single, flexible pull-chain, which can either be dropped down to within easy reach from the ground; or, for a more deluxe application, run into the house. The hook-up uses only one cable, greatly simplifying the problem of turning corners. A spring provides rotation in the opposite direction. See Fig. 6-22 for the overall hook-up.

Note that the antenna mast rides on a thrust bearing base, of the type which is used with electrical TV antenna rotators to help carry the load of a heavy TV antenna. The mast is also supported vertically by a simple wooden bearing.

The cable for operating the rotator can be any small, strong, flexible cable. Stainless-steel aircraft control cable is ideal, and can often be found at bargain prices at surplus stores.

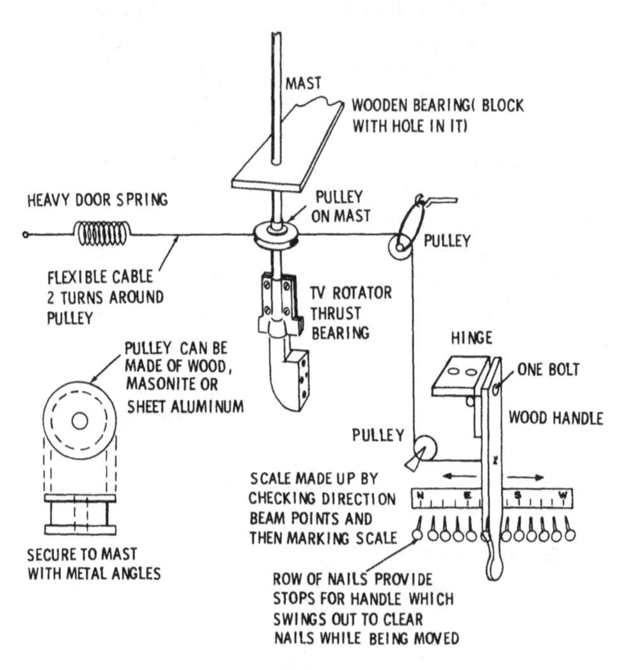

Fig. 6-22. Mechanical antenna rotator.

COMPACTING THE 8JK

The indoor 8JK beam described in Chapter 4 is ideal if you have the space for it. If you lengthen the overall antenna 5%, you can have some bends in the radiating elements which may enable you to mount the antenna in another area which would otherwise be too tight. The bends should be balanced, and will cause less loss if near the end of the radiators (Fig. 6-10).

SNOW ON THE ROOF?

In on-the-air contacts using indoor antennas, the author has had many hams ask, "But what if it snows. Will an indoor antenna work then?"

Experience indicates that snow makes no difference at all; probably because snow, like rainwater, is free of impurities, and therefore does not conduct current readily. At any rate, W0LBV has worked Hawaii many, many times right through a foot of snow.

SUPER-SNEAKY GROUND PLANE

The ground plane described earlier is certainly no eyesore, since the bulkiest part, the radiator, can either be placed on or under the roof. However, the technique can be carried still further, and a mobile auto antenna used for the radiator. These antennas are available for 40-20-15 and 10 meters—wound on *fiberglas* forms.

Three-band mobile antennas are also efficient, and make possible a three-band ground plane. The only change in the layout over the regular three-band ground plane described in Chapter 2 is that the antenna is shorter.

INVISIBLE ANTENNAS

A number of hams have reported success with invisible antennas, used in situations where *any* type of visible antenna was unacceptable. The antennas are simply single wires, end fed with a tuner similar to that illustrated in Fig. 6-7. The secret is the use of #28 or smaller wire, which is very difficult to see once it is in the air. For insulators, small bits of clear plastic—and even button—have been used.

Naturally, such antennas are pretty fragile things, although the wind resistance is so small that they will withstand more abuse than you would anticipate. Certainly not the best answer, nevertheless, such an antenna might be the way you could get on the air in an impossible situation.

CHAPTER 7

Long-Wire Beams

For those lucky hams with an acre or more of ground, or even a large city lot, long-wire beams are highly recommended. Easy to build and easy to tune, such antennas yield outstanding signals on transmit. On receive they will often pull in weak foreign stations which somehow don't seem to come in at all on ordinary beams. Just why the long-wire beams often excel on receive is difficult to explain technically. It may be that the overall length tends to level out fading effects. Perhaps as good an explanation as any is one which an old timer gave the author: "Long wire antennas get a better grip on the ether." At any rate, long-wire antennas work, and they work well.

SINGLE LONG WIRE

A very simple and practical long-wire beam is a single-wire, end-fed antenna. Fig. 7-1 shows the radiation pattern of such a

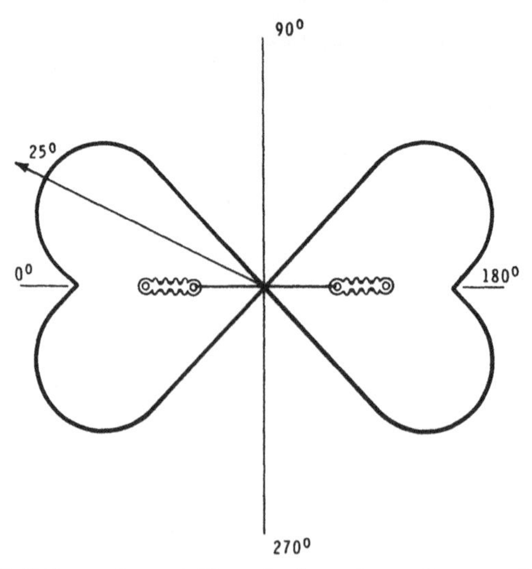

Fig. 7-1. Radiation pattern at 9° angle for antenna three-wavelengths long.

wire 3-wavelengths long. The pattern shown is the one achieved at 9°, a radiation angle desirable for DX. Note that maximum radiation falls along lobes at an angle of 25° to the direction of the wire. As the length of a long wire is increased, the radiation lobes hug the wire more and more. Any wire more than 5-wavelengths long has a direction pattern which, for all practical purposes, is the same as the direction of the wire.

The smallest long wire which gives enough directivity to be called a beam is one 3-wavelengths long. Such a wire yields 2.5-db gain, or, is roughly equivalent to doubling your power. Table 7-1 gives the lengths for a 3-wavelength long wire. Pick the closest to your favorite operating frequency.

Table 7-1. Lengths For A 3-Wavelength Long Wire

7 megacycles	14 megacycles	21 megacycles	28 megacycles
7,000 — 418′	14,000 — 209′	21,000 — 139′	28,000 — 104′
7,150 — 417′	14,200 — 206′	21,200 — 137′	29,000 — 101′
7,300 — 416′	14,400 — 203′	21,400 — 135′	30,000 — 97′

Feeding the Long Wire

The long-wire antenna must be either voltage fed on the end, or current fed a quarter wavelength from the end. The latter method is the easiest, as shown in Fig. 7-2. This type of antenna has the disadvantage of being useful on only one band.

A long-wire beam can be utilized for more than one band by feeding it with an open-wire tuned line and a simple tuner. Fig. 7-3 shows a 21-mc long-wire beam which will also work well on 3.5, 7, 14 and 28 mc. On these bands, however, the beam effect is progressively lost as you move lower in frequency. But even so, the antenna is an effective radiator on all the bands mentioned. It can be fed with the tuner described in Chapter 1 and 30-foot open-wire feeders.

THE SMALL VEE BEAM

The gain from a long wire can be more than doubled by adding another wire, and creating a Vee beam. A 101-foot length for 28 mc, combined with 30′ feeders and a tuned line, makes an ideal all-band antenna and can be used without change with the tuner previously described (Fig. 7-4).

Using a Vee beam of this type on 28 mc with a 120-watt phone rig, the author contacted phone stations all over the world. Not only is such an antenna excellent on receive, but on transmit it seems to give more than its theoretical 7 db of gain. The explanation, probably, is that long-wire antennas do not have to be as high as the usual three-element parasitic array (mounted horizontally) in order to give low-angle radiation for DX. This height problem does not apply to the vertical beams, as explained earlier, and is one of the advantages of the verticals.

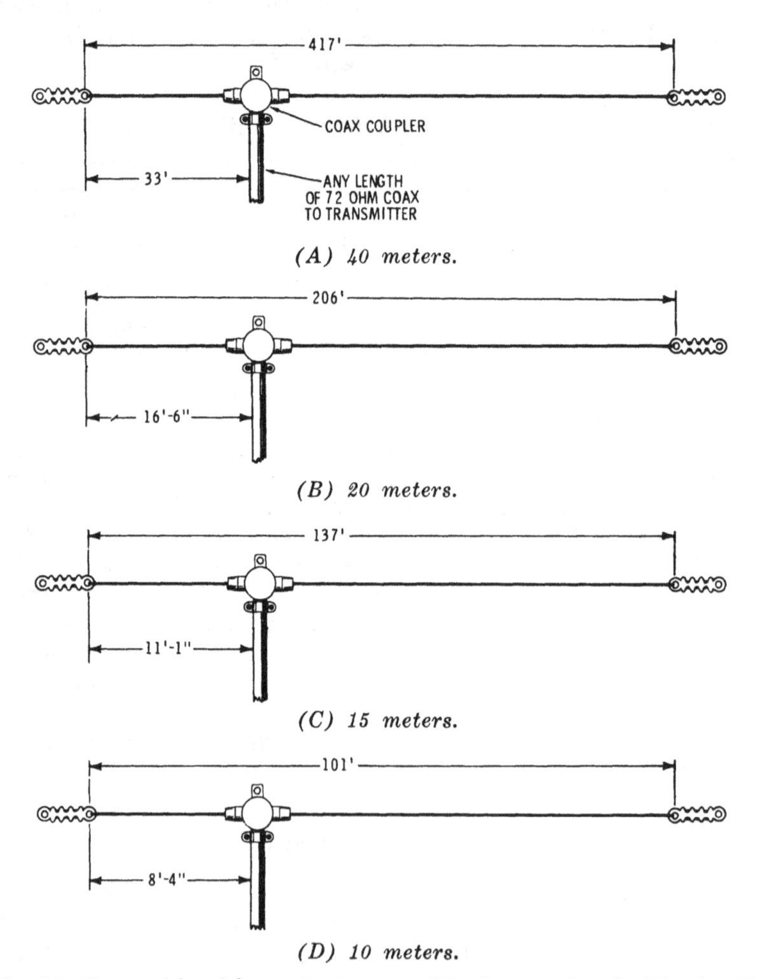

(A) 40 meters.

(B) 20 meters.

(C) 15 meters.

(D) 10 meters.

Fig. 7-2. Three-wavelength-long wire antennas. Gain is approximately 2½ db with a pattern like that in Fig. 7-1.

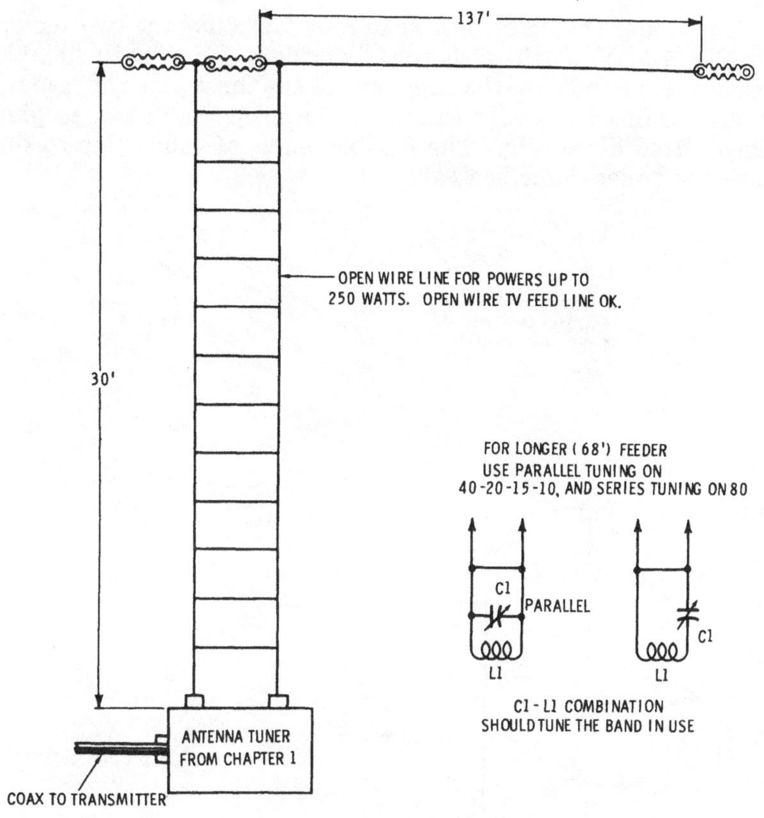

Fig. 7-3. Voltage-fed long-wire antenna for 80, 40, 20, 15, and 10 meters with 2.5-db gain on 15-10.

Obtaining 360° Coverage

Vee beams, like long wires, are bidirectional. As mentioned earlier, a bidirectional antenna properly oriented will give coverage of a surprising portion of the world. However, 360° coverage is desirable if you have the space for the antennas needed to achieve it. By laying out three bidirectional Vee beams, and then providing a way to feed any one of the three, it is possible to achieve around-the-compass coverage (Fig. 7-5).

To feed such an antenna you need a four-wire feeder system. This can be made by peeling spreaders off an open-wire TV line and then using the spreaders (and some household cement) to build a box-like feeder line (Fig. 7-6). Such a feeder will handle up to approximately 250 watts. For high-power, use porcelain spreaders (4″ or so) and No. 14 or No. 12 wire.

The feeder should terminate in four jacks like the two shown in Fig. 7-7. You will also need two plugs fitted with flexible leads and a handle, so that any pair of the three pairs of feeders can be utilized in order to choose the proper antenna to give the desired directivity. The flexible leads, of course, go to the antenna tuner described earlier.

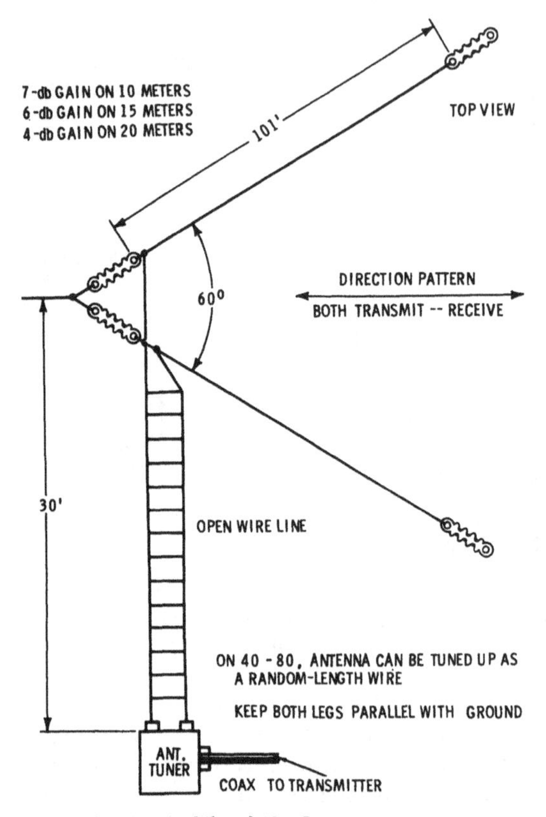

Fig. 7-4. Multiband Vee-Beam antenna.

RHOMBIC ANTENNAS

The sophisticated cousin of the Vee beam, the Rhombic, is actually seldom used by hams because of the considerable space needed for its erection. This is a shame because the rhombic is capable of very high gains. For example, a rhombic which has lengths of three wavelengths on a side gives more gain than a Vee beam six wavelengths on a side. Either antenna is capable of output in the range of 10 db over that of a simple dipole

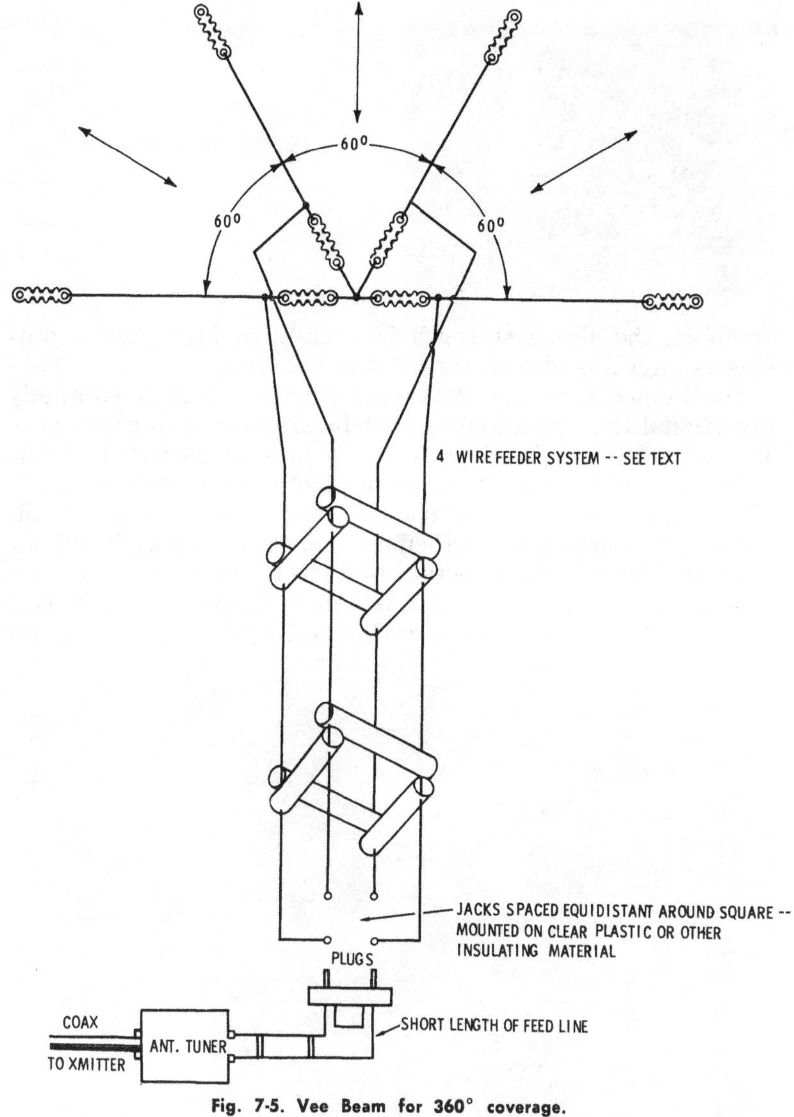

Fig. 7-5. Vee Beam for 360° coverage.

which is equivalent to stepping up the power of your transmitter 10 times.

In actual use, assuming that the rhombic is mounted fairly high above ground, the practical gain achieved in communication seems to exceed even the theoretical gain. As you know, if you have ever heard a station on a ham band using a

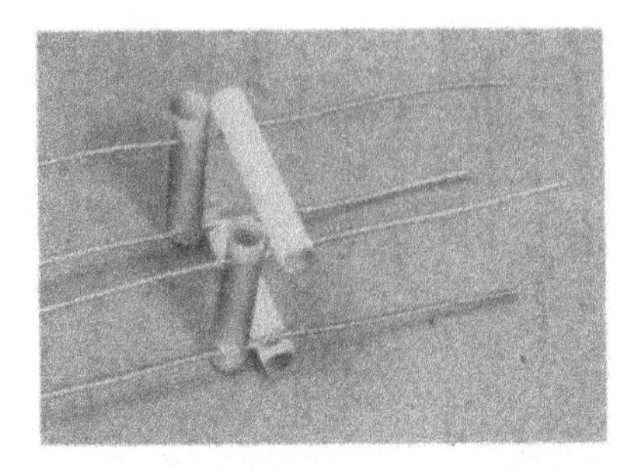

Fig. 7-6. Line construction.

rhombic, the signal strength is usually so high that it outclasses anything else on the band at the time.

On the negative side, the larger rhombics become so highly directional that the antenna must be oriented with great care. In fact, the directional pattern becomes so narrow that the antenna is useful only for point-to-point communication. In addition, the classic type of rhombic transmits only in one direction, this uni-directional effect being achieved by terminating the antenna with some type of non-inductive resistor.

For ham use there is one fairly good answer: the resonant rhombic. Such an antenna is bidirectional, requires no re-

Fig. 7-7. Terminating the line in jacks.

sistor terminating hook-up, and is broad enough in pattern so that aiming it does not necessitate techniques beyond the facilities of the average ham. Fig. 7-8 shows the overall layout of one practical rhombic antenna which is about as big as even the most fortunate hams have space to build. It is designed primarily for the 20-, 15- and 10-meter bands. However, it will put out an effective signal on 40 and 80 meters as well.

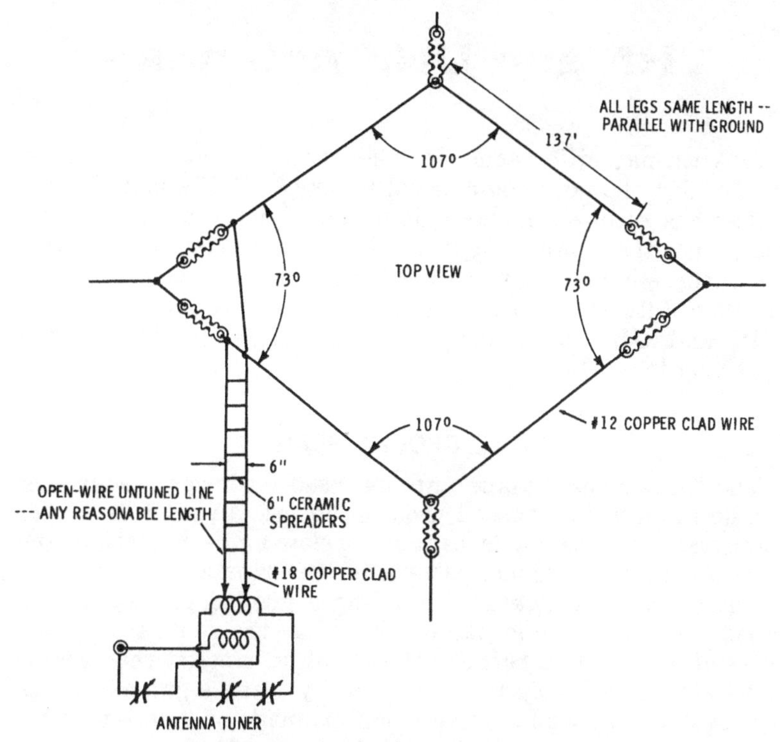

Fig. 7-8. Resonant-rhombic antenna construction.

The antenna is fed with an open-wire line, which in turn is matched to the usual transmitter coaxial-line output by means of an antenna tuner. Fig. 7-8 shows the circuit of the tuner, which can be the one described in Chapter 1. The tune-up technique is covered in detail in Chapter 3.

CHAPTER 8

VHF and UHF Antennas

As with parasitic beams for the lower frequency band, the availability of many commercial types of VHF and UHF antennas has resulted in the wide usage of manufactured types. Also, since the simpler VHF and UHF antennas are relatively small, there is not the same need to work out a custom layout for each different location. Even so, there are basic types of VHF and UHF antennas which can be built easily for a relatively low cost.

50-MC GROUND PLANE

The same ground-plane antenna used on lower frequencies can be reduced in size and used on 50 mc. As with its larger brothers, the antenna is capable of low-angle radiation and gives an omni-directional pattern for covering a full 360°.

There are many ways of building a ground plane. The so-called drooping ground plane, which has the radials drooping down at an angle, is perhaps the simplest, since it requires no elaborate matching system. Further, by using a gamma-match feed system, the whole antenna can be made of metal, without the need of any insulation between the vertical antenna and the radials, greatly simplifying mechanical construction.

Fig. 8-1 shows a drooping ground-plane antenna which is the author's modification of an antenna originally worked out by K4PRL. It is simple to build and requires no hard-to-get parts. The radiating portion of the antenna is actually the top half of a 10-foot aluminum TV-antenna mast. The point at which the radials are attached establishes a ground. The portion of the antenna above that point is the actual radiator. The radials are slightly longer than the radiator.

Notice that TV U-bolt assemblies are used to mount the various components. One connects the top end of the gamma-matching section to the TV radiating mast. The other end of

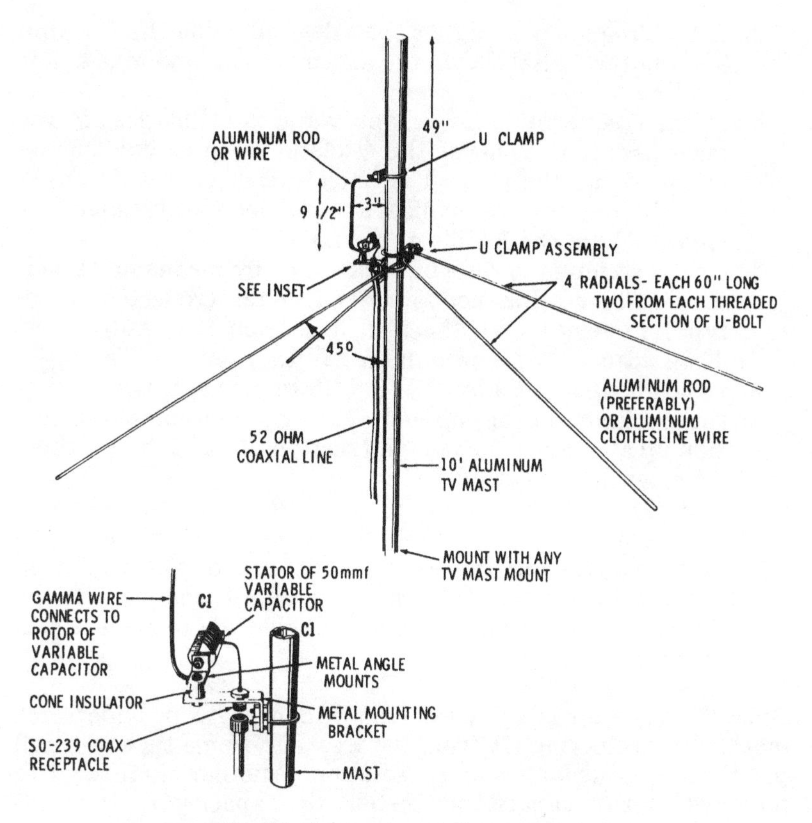

Fig. 8-1. 6-meter ground-plane antenna.

the gamma section goes through a variable capacitor to the hot terminal on an SO-239 coaxial fitting, which is mounted on an angle that in turn goes to a *U* bolt. Another *U* bolt serves to mount the radials, which ideally should be made of stiff aluminum rod or tubing, although aluminum clothesline wire can be used if nothing else is available.

Tuning Up With Instruments

Once the antenna is assembled as shown in the drawing, here is the tune-up procedure:

1. If you have a grid-dip oscillator available, the first step is to resonate the antenna to the desired frequency. With the antenna set up some place where you can reach it, couple the GDO to the center of the antenna and check the frequency.

2. If the frequency is higher than desired, slide the *U* clamp holding the radials down the antenna mast and take a new reading.
3. If the frequency is lower, you want to utilize less of the mast as antenna, hence the *U* clamp with radials should be moved up the mast. The objective, of course, is to get the antenna on resonance at the desired frequency—frequently in the middle of the band.
4. The next job is to tune out reactance by means of capacitor C1. To do this, hook an SWR meter (reflected-power type) in series with the feed line from the transmitter. Then adjust C1 for minimum SWR.
5. If you cannot get a low SWR (1.5 to 1 or so), try moving the *U* clamp on the top end of the gamma-matching section up and down the mast a fraction of an inch at a time. After each move adjust C1.
6. Ordinarily, this technique, particularly if the antenna was resonated with a GDO, will give a good SWR. However, readjusting the electrical length of the radiating portion of the mast by making slight changes in the position of the top *U* clamp may enable you to achieve still lower SWR.

Once the tune-up is achieved, you will need to provide some means for protecting C1 from the weather. Some type of small plastic box is usually a good solution. Another method is to measure (with a capacitance meter) the capacity of C1 as adjusted for best match and then substitute a fixed mica capacitor of that value. This, then, can be covered with several layers of electricians tape and sprayed with a TV insulating spray.

Tuning Up Without Instruments

Tune-up can be done fairly well without instruments, although results will not be as accurate as the method just outlined.

1. Set up the antenna exactly to the dimensions shown in Fig. 8-1.
2. Carefully adjust C1 for maximum load on the transmitter.
3. Make a very minor adjustment (say, ½″) in the location of the gamma-match connection to the mast. Again check for maximum load. Proceed in this fashion—but in no case settle for an adjustment which is greatly different from the dimensions shown. If you seem to be getting such results, use an SWR meter.

A GROUND PLANE FOR 144 and 220 MC

This antenna should be used only where it can be fed with a fairly short length of coaxial cable; losses become extremely high at these frequencies. Anything over approximately 50 feet can introduce serious loss on 220 mc, even with a very low SWR. The problem is not quite so critical on 144 mc though.

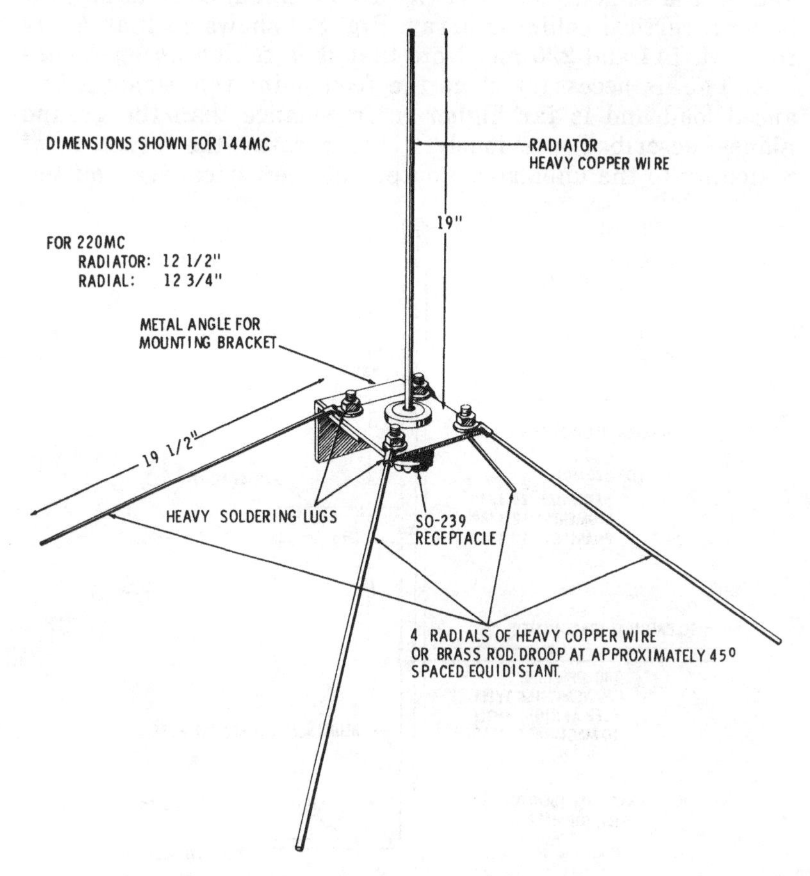

Fig. 8-2. 144- and 220-mc ground-plane antenna.

Fig. 8-2 shows the layout of the antenna. The elements should be made of some type of stiff copper wire or 1/8" brass rod; for example, the type used in welding. Since there is no easy way to telescope elements, probably the best way to get the antenna to the exact resonant length is to cut the radiator a bit long, and then cut it back in length, as indicated by a field-strength meter, or by the *S*-meter on a receiver. However, the antenna should perform fairly well is simply cut to dimension.

360° PATTERN WITH BEAM EFFECT

All of the antennas described so far give vertical polarization and yield no gain. It is possible to achieve a 360° pattern and, at the same time, boost the usable signal considerably by using a vertical collinear array. Fig. 8-3 shows such an array for both 144 and 220 mc. Note that it is fed by an open-wire line. This is necessary since the feed point represents a balanced load and is far higher in impedance than the ground planes described previously. This creates the problem of matching to the unbalanced output of the typical transmitter.

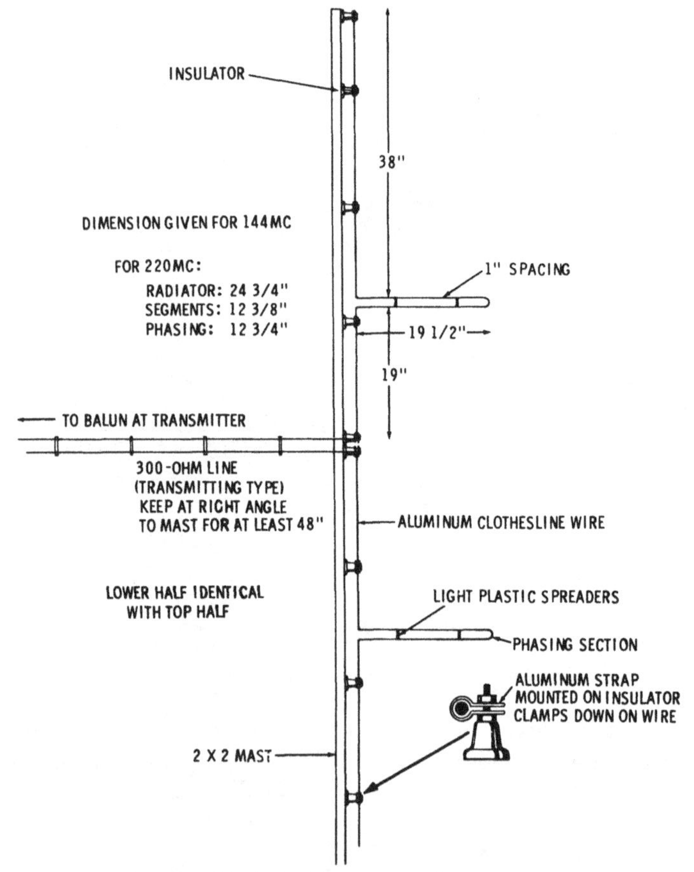

Fig. 8-3. Vertical collinear array for 144-220 mc.

As was covered in detail in Chapter 1, one answer is to use a balun. However, instead of using coils, as was the case at the lower frequencies, the balun is constructed of short lengths of coaxial cable. This permits use of relatively low-loss open wire on TV transmitting-type twin lead from the antenna to the operating point, and then dropping the impedance to the proper point there with the balun. The balun coax can be coiled up inside a small chassis in order to make a neat layout (Fig. 8-5).

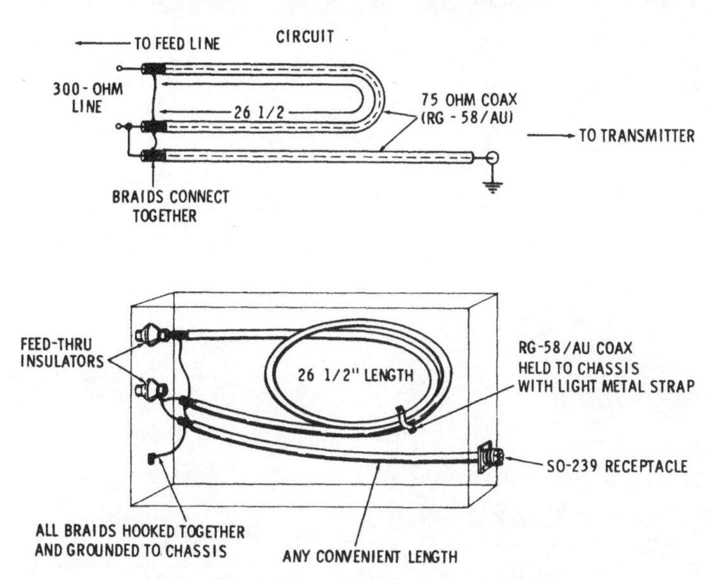

Fig. 8-4. Balun made from coaxial cable.

An antenna of this type requires some sort of wooden pole to support it. A 2 × 2 board is strong enough, if you are careful to select one free from knots. As shown in Fig. 8-3, the antenna is supported by means of stand-off insulators and light metal straps using the technique illustrated a number of times earlier in the book. Actually, at the low-voltage points, even TV stand-off insulators will give adequate insulation. However, since more insulation is necessary at other points along the antenna, you might as well use a good grade of ceramic stand-off in the interest of uniformity.

No tuning adjustment is provided with this antenna. Simply hook the feeder to the proper point, run the 300-ohm line to the transmitter, and use a balun in the line to get the proper impedance match.

ACHIEVING 360° COVERAGE WITH HORIZONTAL POLARIZATION

Vertically polarized antennas are frequently used on 144 and 220 mc, particularly for communicating with mobile rigs. However, most long-distance communications at these frequencies takes place with horizontally polarized antennas, usually beams. A horizonally-polarized antenna which gives 360° coverage without having to be rotated is most useful.

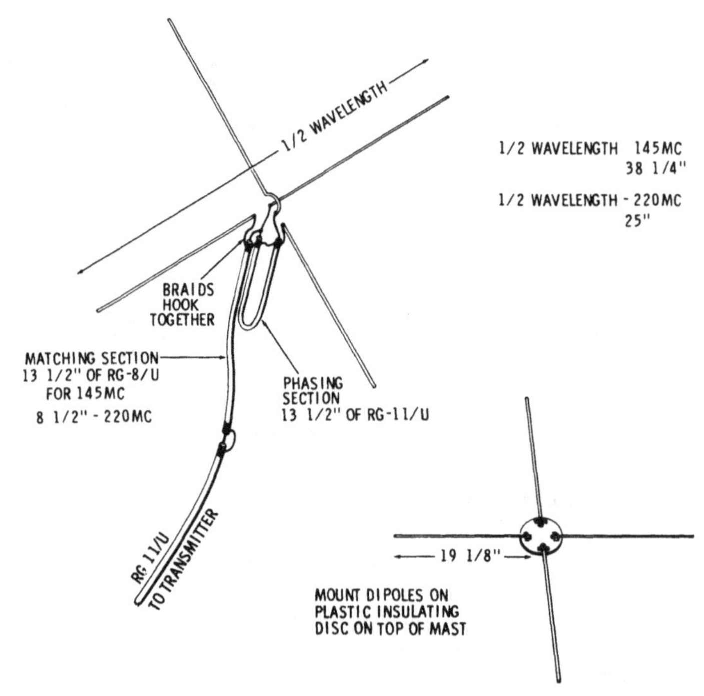

Fig. 8-5. Turnstile antenna.

Fig. 8-5 shows such a hook-up, called a turnstile antenna. As will be seen, it consists of two dipoles that are crossed at right angles to each other and fed 90° out of phase.

With the antenna, both a phasing and *Q* section are provided, the first to get the desired phase shift, and the second to achieve a better match with the feed line. No tune-up should be required. Simply cut the antenna to length following Fig. 8-5.

SIMPLE 144-MC BEAM

Even a small beam makes such a big improvement in performance of both transmitter and receiver on 144 mc that you want to have one if it is at all possible. Here is a beam so simple that it can be built in an hour or less. It is quite small, too, and therefore can even be used as an indoor antenna.

The layout is shown in Fig. 8-6. Here, copper tubing can be used for the elements. However, the element material from a

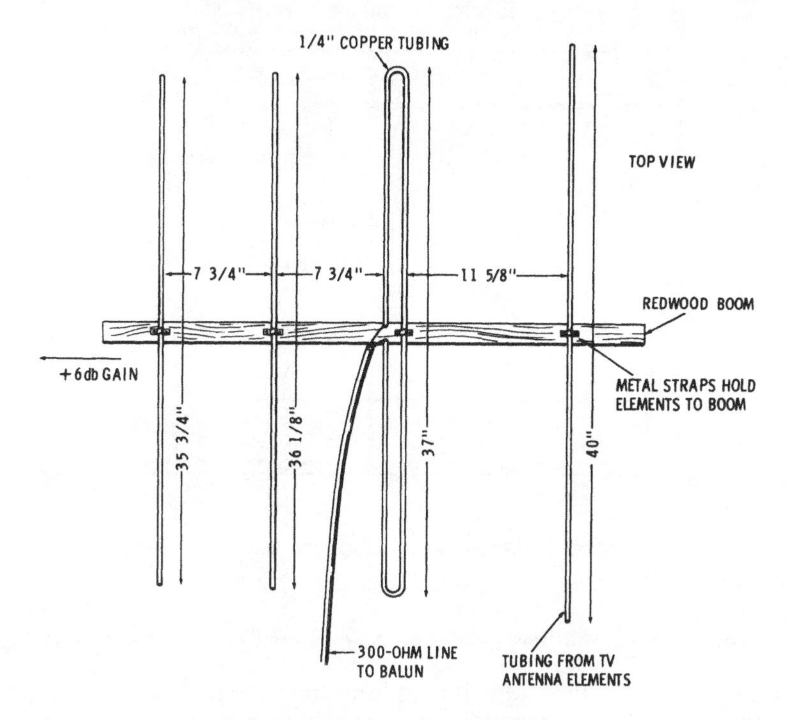

Fig. 8-6. Compact 2-meter yagi beam.

TV antenna is ideal for everything except the folded dipole. For this, copper tubing is easier to bend; use the kind of soft tubing used for auto gas lines.

For easy construction you can make up the elements first, and then secure them to the wooden boom. Redwood is ideal for the latter, since it resists rot better than most other woods. You can paint it with aluminum paint so that it matches the elements in appearance.

If you want to avoid the problem of having to bend the tubing, you can use a single-element radiator, as shown in Fig. 8-7 and feed it directly with the 300-ohm line. The antenna will be somewhat more sharp-tuning than with the folded dipole.

The taps have to be adjusted by experiment, in which case you use the element spacing shown in Fig. 8-6. However, if you will widen out the spacing, tuning will be less critical and the dimensions shown for the taps in Fig. 8-7 should be close to

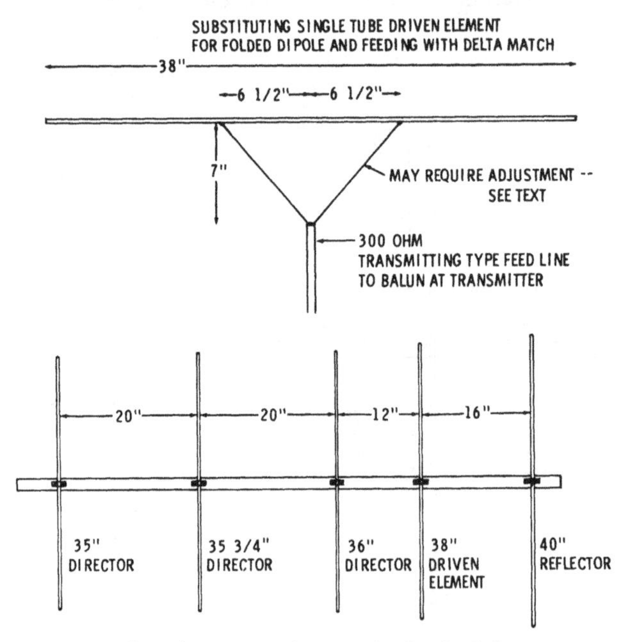

Fig. 8-7. Array with wider spacing for a wider band of frequency coverage.

being correct. The addition of another element will increase gain somewhat, achieving, in total, a 6 to 8 db increase in forward direction over a dipole. This is the equivalent of raising your power by at least 4 times.

LONG WIRES FOR VHF

The long-wire antennas, particularly the rhombic, *can* be used on the VHF bands. Here are the dimensions for a rhombic which will work on both 2 and 1½ meters, and will yield excellent TV reception on channels 7 through 13 as well. Notice in Fig. 8-8 that the antenna is terminated with resistors, which makes the antenna unidirectional in the direction toward the resistors—which must be non-inductive (not wirewound) carbon types. A pair of 2-watt resistors will handle up to a 10-watt transmitter. Beyond that it is necessary to make up a series parallel hook-up with the resistors.

For high-power, the best way to dissipate the power is to use a long, open-wire line, deliberately made up to have considerable resistance. Such a line should be at least 6-wavelengths long. It can be made up with 2" spreaders and any stainless-

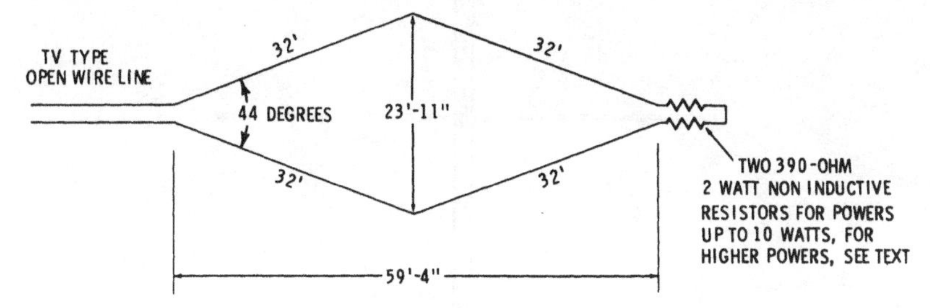

Fig. 8-8. Top view of a VHF rhombic antenna.

steel wire which is roughly the equivalent to number 26 copper wire in diameter. The line should finally be terminated by 2-watt, 390-ohm resistors.

UTILIZING TV-ANTENNA PARTS

In the antennas already described, some use was made of TV antenna parts. Actually, you can go all the way and use TV antennas for everything. In fact, some of the simpler wide-band TV antennas (for example, the popular conical) will work after a fashion on 50, 144, and 220 mc. Of course, an

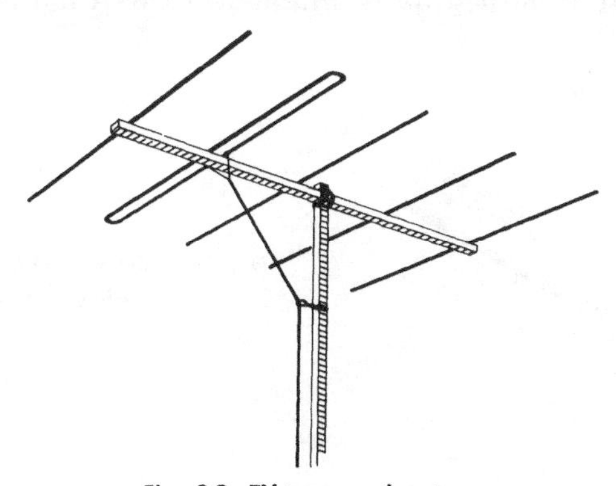

Fig. 8-9. TV-type yagi antenna.

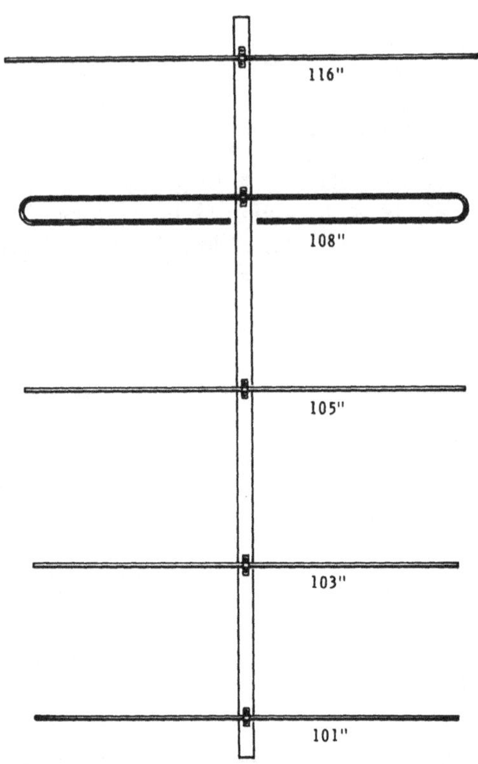

Fig. 8-10. Suggested lengths for modifying Channel-2 TV yagi for 6 meters.

antenna cut to a specific band of frequencies works far better. It happens that a yagi-type TV antenna (Fig. 8-9) comes in two sizes which are very close to the ham bands. One of these is the antenna for Channel 2. Actually, the 6-meter ham band includes what was originally intended to be Channel 1.

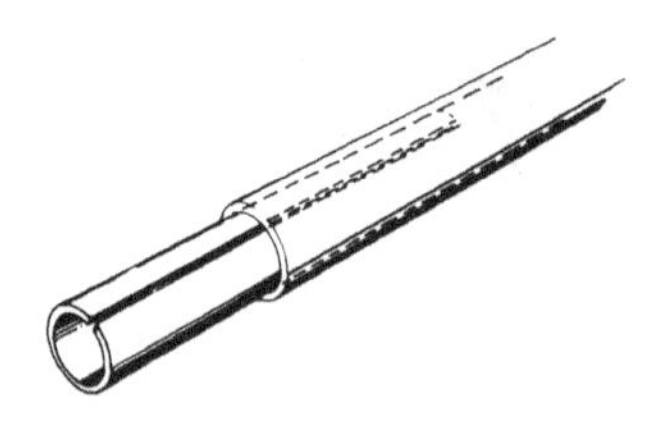

Fig. 8-11 Slotting tubing will allow adding inserts.

Six-Meter Yagi

Since Channel 2 is higher in frequency than the ham band, all of the elements of the antenna are a bit short and had best be made longer following the dimensions given in Fig. 8-10. You can lengthen the elements by cutting off the crimped ends and then very carefully adding short pieces which have been slit as shown in Fig. 8-11. The tubing can be acquired by removing the second director and making a four-element array. Another way to get tubing is to buy a stacking kit, which usually has tubing the same size as the elements on the TV antenna.

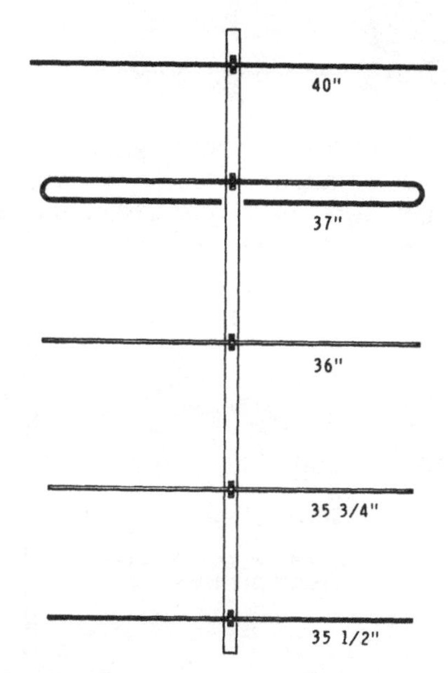

Fig. 8-12. Suggested dimensions for cutting down Channel-5 TV antenna for use on 144 megacycles.

A four- or five-element yagi made up in this fashion is reasonably broad tuning so that no elaborate tuning method is required. You will need a balun to connect the feed line to the transmitter. If you have the equipment, you can adjust the element length for maximum output as indicated on a field-strength meter, or on the *S*-meter of the receiver of one of your ham friends.

Two-Meter Yagi

A two-meter yagi is a lot easier to build from a TV yagi since, in this case, a TV antenna is available which has longer elements than needed, and there is nothing to do but cut them

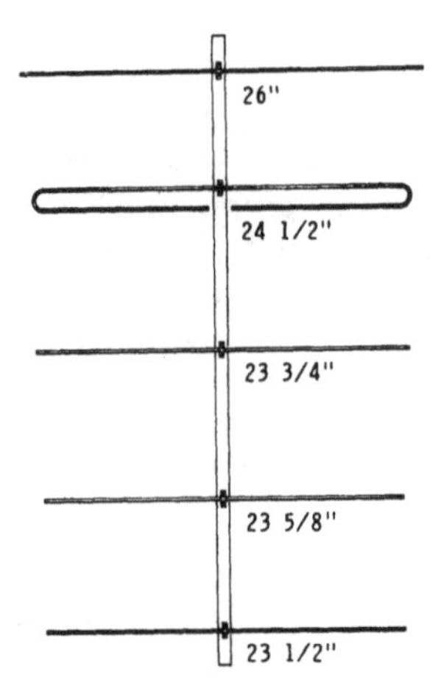

Fig. 8-13. Cutting down Channel-13 yagi for use on 220 megacycles.

to the proper length as indicated in Fig. 8-12. Channel 5 antennas are the closest to being the right length. However, the longer Channel 4 or Channel 3 antennas will do just as well. As before, the antenna can be fed with a 300-ohm line, preferably

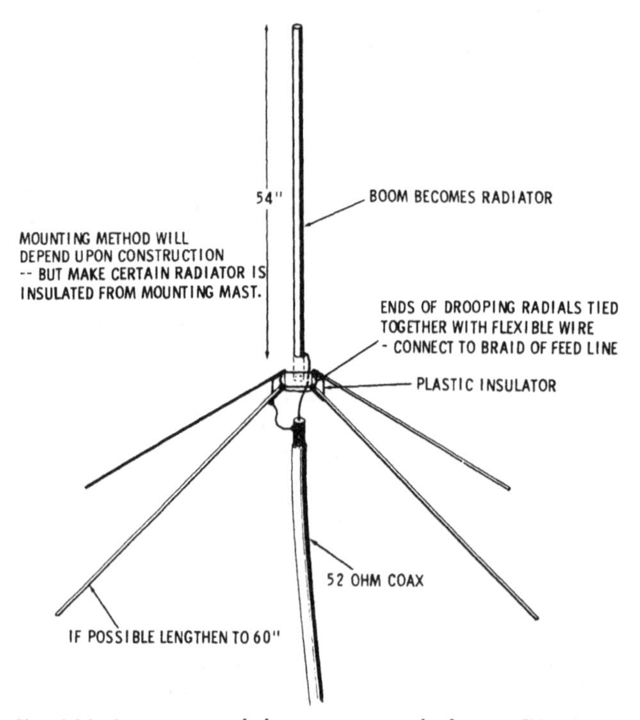

Fig. 8-14. 6-meter ground-plane antenna made from a TV antenna.

transmitting type, and hooked to the transmitter by means of a balun.

1½-Meter Yagi

TV yagi antennas in the 7-13 Channel range can likewise be modified very easily to provide 220-mc antennas. Fig. 8-13 shows the lengths of the elements.

Five-Meter Ground Plane

The ground plane described previously uses an all-metal type of design, an advantage in home-brewed gear because providing adequate insulation can be quite a problem. However, one good answer is to use a conical TV antenna on end. The boom becomes the radiator, and the elements, which were once the antenna, provide the drooping radials.

The radials should be lengthened by slotting left-over tubing from the other element on the original TV antenna. Likewise, the TV mast should be lengthened to provide the proper resonant length by adding a piece of tubing. Put some self-tapping screws into the tubing to join the pieces together firmly. Fig. 8-14 shows the overall layout.

CHAPTER 9

Special-Purpose Antennas

Chapter 1 covered several types of tuned antennas which will operate efficiently over a wide range of frequencies. The antennas were all end-fed. With most ham layouts this is the more convenient physical arrangement, since the ordinary operating point is at one end or other of the antenna. However, if you are set up so that tuned feeders can be run to the center of the antenna, the tuned doublet has some inherent advantages which make it deserve consideration.

TUNED DOUBLET

Fig. 9-1 shows a typical tuned doublet. Ideally, the antenna would have an overall length of 134 feet, in which case it would be an effective radiator on all bands from 80 to 10 meters. Actually, it can also be used on the 160-meter band (a band which has seen little use in recent years but is undergoing something of a comeback because it is more effective than 80 meters for short-haul mobile work).

The tuned doublet can be fed very readily with the tuner described in Chapter 1. Different lengths of antennas and feeders require different tuning methods; you will have to adapt the plug-in coil arrangement to provide either parallel of series tuning as indicated in Table 9-1.

To tune up the doublet with an antenna tuner, follow the detailed instructions in Chapter 3.

The tuned doublet has an advantage over the end-fed antenna with tuned feeders in that it is a balanced system with considerably less danger of unwanted radiation from the feeders. Also, on all bands for which it is used, it can be brought to resonance by tuning adjustments of the tuner. On the negative side, in addition to the frequent problem of the feeders not being where you would like to have them for easy feeding from the operating point, the radiation pattern of the doublet changes radically as you move from band to band. For example, the 135-foot antenna functions as a simple dipole on 80 meters, with the *figure-8* pattern characteristic of the dipole. On 40 meters, the antenna becomes two half waves in phase, with a rather sharp *figure-8* pattern, which means that it will be difficult to work stations off the ends of the antenna. On 20 meters, the pattern breaks into a number of lobes, which tend

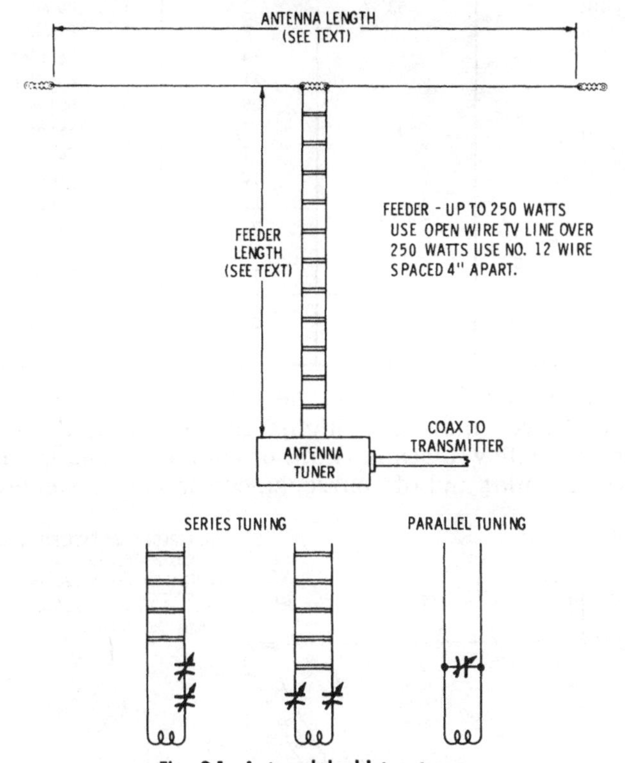

Fig. 9-1. A tuned-doublet antenna.

to give better 360° coverage. On 15 and 10 meters there are so many lobes that there is no difficulty, usually, from excessive directivity.

360° Coverage

There is one way to handle the doublet which does insure 360° coverage on the next higher band than that for which the antenna is cut. This is accomplished by mounting the two halves of the antenna at right angles to each other. This will, of course, detract somewhat from the radiation on the funda-

Table 9-1. Antenna and Feeder-Length Tuning Methods

Antenna Length	Feeder	Band	Type of Tuning
135′	42′	1.7	Series
		3.5	Parallel
		7	Parallel
		14	Parallel
		21	Parallel
		28	Series
105′	83′	3.5	Parallel
		7	Parallel
		14	Parallel
		21	Parallel
		28	Series
67′	33′	3.5	Series
		7	Parallel
		14	Parallel
		21	Parallel
		28	Parallel
44′	33′	7	Parallel
		14	Parallel
		21	Series
		28	Parallel or Series

mental frequency; but even so the antenna will be an effective radiator. As you move still higher in frequency, the antenna becomes a small Vee beam. Gain, of course, is small, but even so is worthwhile; and of course, since the gain is achieved by

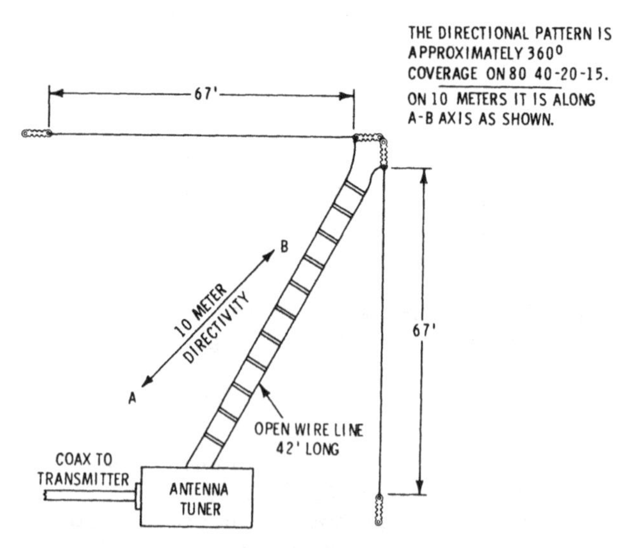

Fig. 9-2. 360° coverage with a horizontal multiband antenna (top view).

added directivity, this must be taken into consideration in orienting the antenna. Fig. 9-2 shows a typical antenna designed for 80- to 10-meter operation. It will give a satisfactory around-the-compass pattern on 80, 40 and 20 meters; some directivity on 15 meters, and sufficient (3 db) gain on 10 meters to provide a worthwhile increase in both effective power output and in the low-angle radiation so desirable for 10 meters.

Improving the Tuned Doublet

The impedance which the doublet presents to the feed system varies over quite a wide range in changing bands. This is undesirable on two counts: voltage may become quite high at several points on the feed line, and, in addition, tuning may become pretty critical.

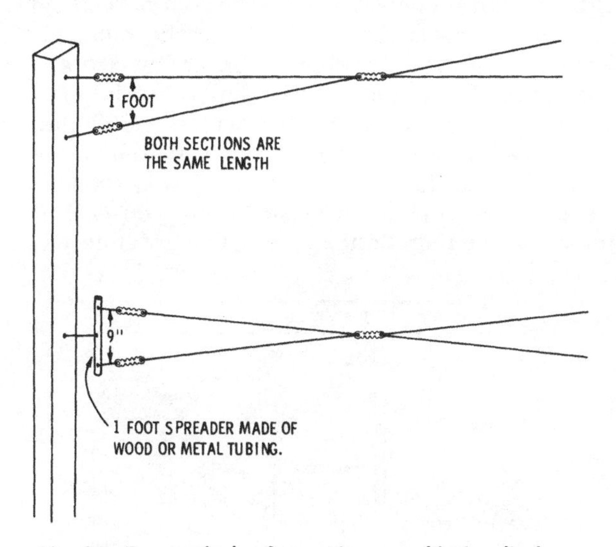

Fig. 9-3. Two methods of mounting a multi-wire dipole.

A very worthwhile improvement can be made in the tuned doublet by fanning out the ends. Fig. 9-3 shows two different ways of doing this. One involves using two wires for each portion of the doublet and simply running them to the support mast. The second method is often easier to erect in that the pull can be applied with only one haliyard from the mast. Note that the pull is applied off center. This is done to make the antenna wires self-tightening in the sense that the same tension is automatically applied to both wires.

THE WINDOM

The widespread use of multiple-band, bandswitching transmitters in recent years has greatly increased interest in antennas which can be used on a number of bands without any type of tune-up in going from one band to the other. All such antennas entail some compromise in design, even in the case of the triple-band type of parasitic beam, into which a great deal of engineering has been put in an effort to get an efficient three-band system. And the five-band trap-type doublets generally are not as efficient as the three-band beams. Each band which is added increases the technical problems involved, and the user has to accept some loss in order to enjoy the convenience of not requiring a tuner.

If you are willing to accept some inefficiency, there are home-made types of antennas, as well as the commercial types, which will give multiband coverage. Perhaps the most popular is a type usually called the Windom, or off-center-fed antenna. The theory in back of this antenna is that there is one point on a harmonically operated antenna at which a 300-ohm feed line can be attached with a fair impedance match. Unfortunately, the technical facts do not fully support this theory. However, the antenna does work satisfactorily and you will find a lot of them in use on the ham bands today. The antenna will cover 80,

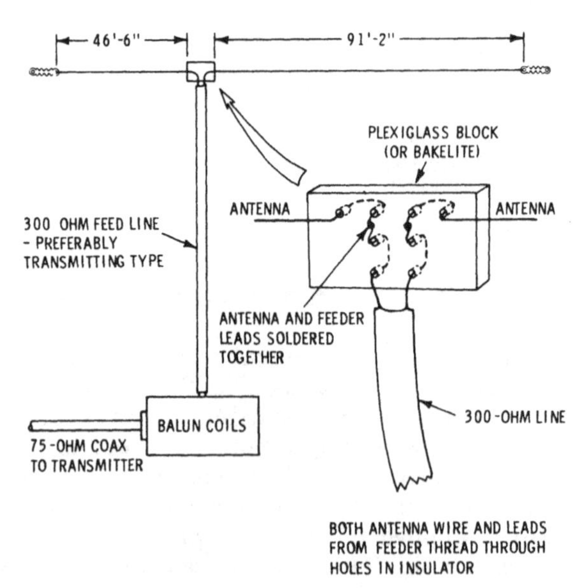

Fig. 9-4. The Windom antenna for 80, 40, 20, and 10 meters.

40, 20 and 10 meters. (It will not work well on 15 meters because that band is not harmonically related to the others in the proper way.)

Fig. 9-4 shows the hook-up of the antenna. Some of its users insist that the design of the center insulator is quite important —the idea being to avoid any fanning of the feed line at the point of attachment. Whether or not this is a critical matter is a moot question. At any rate, the method shown makes a neat installation. Since the antenna has a 300-ohm feed line, it is necessary to use the balun coils described in Chapter 1 to provide an unbalanced input to match to the transmitter. Note that the coaxial line is 72 ohms. The use of the balun also tends to keep stray RF away from the transmitter, particularly if the balun is located some distance away from the transmitter.

It is important with an antenna of this type to keep interaction between the antenna and the feed line to a minimum, which means that the feed line should be kept at right angles to the antenna for as far as possible. The length of the feed line is not critical, ordinarily, although if you have difficulty getting the antenna to load on one or more bands, experiment with different lengths.

THE LATTIN 5-BAND ANTENNA

The Windom antenna just described uses the entire antenna length on all the bands on which it is operated. This is a definite advantage, and one which somewhat tends to compensate for the rather dubious impedance match. The antenna which follows operates on an entirely different principle—that of providing trap circuits which, in effect, reduce the antenna to a doublet on the band in use. For example, when the antenna shown in Fig. 9-5 is used on 10 meters, only the center 16 feet actually radiates.

The Lattin antenna, named for the ham (W4JRW) who invented it and holds a patent on the basic principle, utilizes quarter-wave stubs which, at the frequency to which they are cut, act exactly as if they were an insulator. For example, in Fig. 9-5, the 6′-11″ stub (quarter wave times the velocity factor .8 of the feed line used) blocks the RF for 28 mc from reaching the rest of the antenna. Many commercial antennas use parallel-tuned trap circuits for exactly the same purpose. The Lattin technique has the advantage of being much easier (and less expensive) to build. The whole antenna is made up of a length of transmitting-type 300-ohm twinlead, the tubular type filled with a plastic foam to exclude moisture.

Fig. 9-5 shows the mechanical construction of the antenna. Building it will require some patience, and a sharp (preferably brand new) pair of side-cutting pliers. The pliers are necessary for cutting one side of the twin-lead and stripping off the insulation to allow for making the stubs. Be very careful not to cut the top lead—remember, it has to carry the full weight of the antenna.

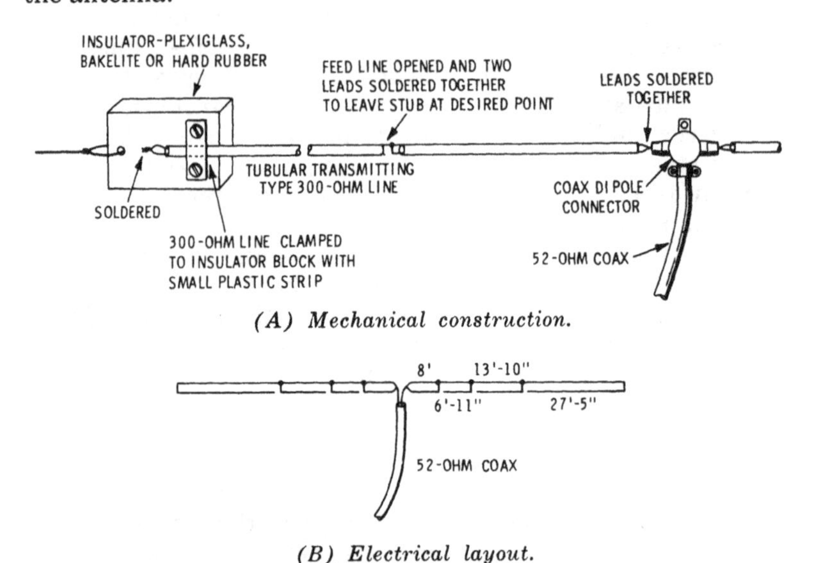

(A) Mechanical construction.

(B) Electrical layout.

Fig. 9-5. Lattin 5-band antenna for 80, 40, 20, 15, and 10 meters.

After the stub connections have been made, the next step is to spray the connection with TV high-voltage spray and then carefully tape the junction. The spraying should include the open stub ends, which are a point of high voltage when the band for that stub is in use by the transmitter. The insulators can be fabricated from hard rubber or clear plastic.

FOR THE NOVICE: A 3.5-7-21 MC ANTENNA

There is another technique for building a multiband antenna which lends itself especially well to the three Novice bands. Actually, it consists of an 80-meter doublet with a 40-meter doublet hung directly below it. Since a 40-meter doublet will also function as a 3/2-wave antenna on 15 meters, the resulting antenna gives three-band operation.

Fig. 9-6 shows such an antenna. One very easy way to build it is to construct it entirely from open-wire TV line. The top wire of the two-wire feeder provides the radiator for 80 meters. The lower wire is cut off at the proper point to furnish a 40-meter dipole. An antenna of this type will catch a certain amount of wind, so, as a precaution add some household cement to the points at which the plastic spreaders are affixed to the feed wires. Also, allow the end of the 40-meter dipole to extend half an inch or so beyond the spreader insulator. Secure the dipole wire to the insulator by wrapping the junction with fine wire (about No. 30) and then coating the joint with household or china cement.

Although not absolutely necessary, a regular coax coupler is advisable for the center of the antenna. Both wires on each side can be put into the coupler, which will provide both a good mechanical and a good electrical joint. The coax line to the transmitter can be any reasonable length.

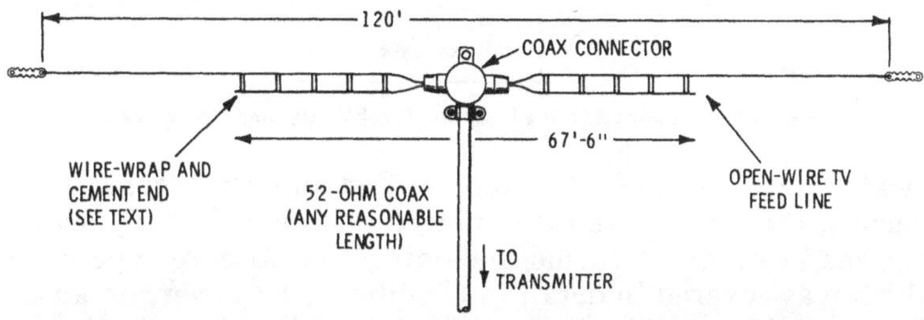

Fig. 9-6. Novice-band antenna for 80, 40, and 15 meters.

COMPACT FOLDED DIPOLE FOR THREE NOVICE BANDS

As has been pointed out previously, it is possible to use a 40-meter dipole on 15 meters. In addition, it is also possible, by taking advantage of the matching characteristics of a quarter-wave line, to use the 40-meter dipole on the 80-meter Novice band. This allows an antenna with an overall length of less than 66 feet—an important consideration for the ham with limited space.

Fig. 9-7 shows the antenna and the proper dimensions. It can be fed from a Balun as described in Chapter 1; however, a tuner of some type is a better idea, since this antenna, like most other multiband antennas, does not discriminate against harmonics.

END-LOADED NOVICE-BAND ANTENNA

Just as was the case with the compact beam antenna described earlier, loading coils can be used to piece out the elec-

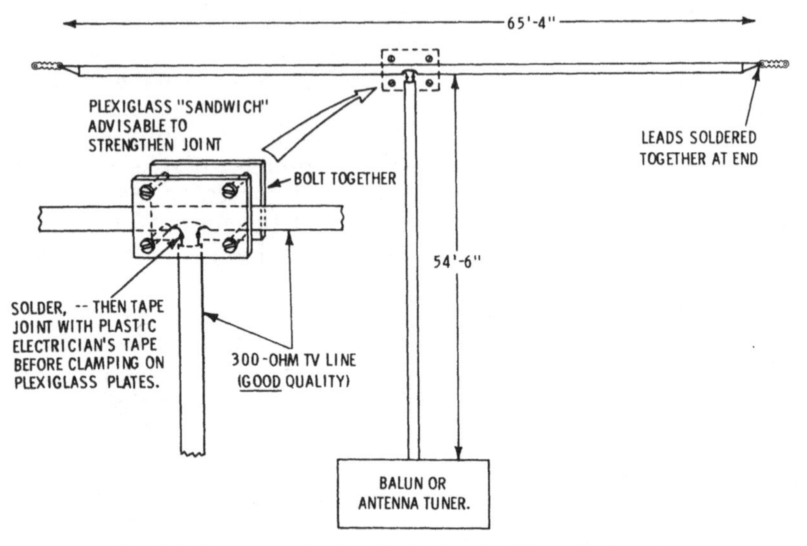

Fig. 9-7. A compact folded dipole for 80, 40, and 15 meters.

trical length of a wire antenna. For practical construction reasons, these coils are often placed on the end of dipoles and take many different forms, including the air-core type of coil which was covered in detail in Chapter 1. However, because of the inevitable wind resistance, loading coils of small diameter

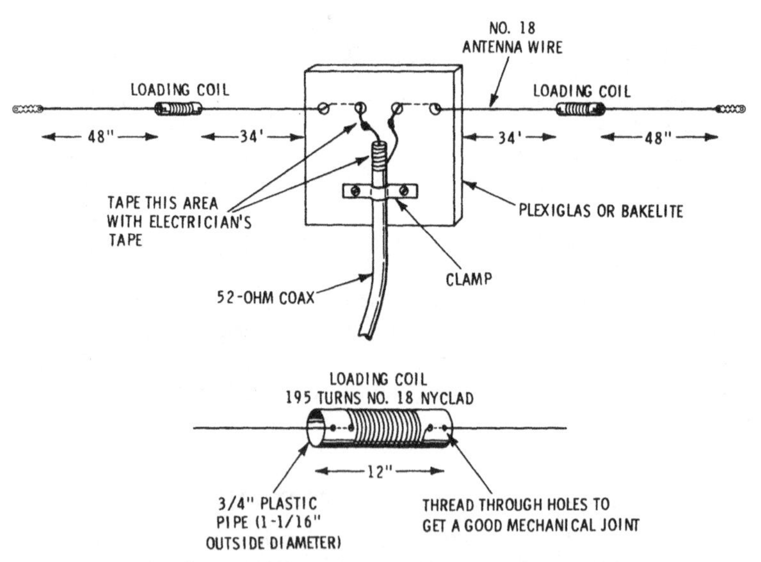

Fig. 9-8. 3-band antenna for 80, 40, and 15 meters using loading coils.

are desirable. Fig. 9-8 shows an antenna which uses loading coils of this type.

The coils each consist of 195 turns of number 18 Nyclad wire close wound on ¾″ plastic water pipe. This pipe has an outside diameter of 1-1/16 inches. The plastic pipe is obtainable at many hardware stores and at most firms handling pumps. It is both inexpensive and a good insulator.

You can adjust the antenna to exact resonance by lengthening or shortening the 48″ portion to achieve lowest SWR on 80 meters and by doing the same thing with the longer section on 40 meters. The usual coaxial connector is recommended for the center of the antenna.

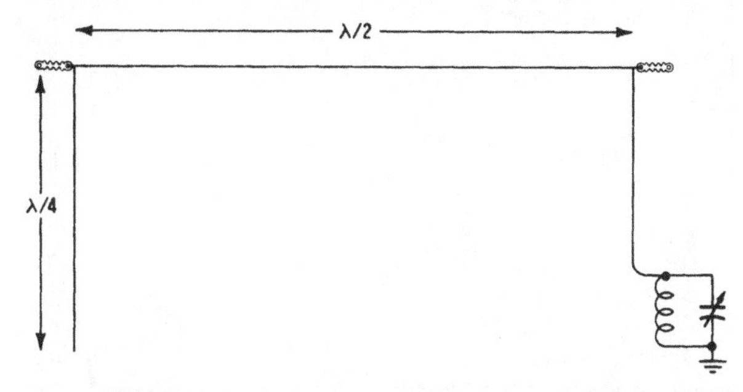

Fig. 9-9. The upside-down vertical for 7- and 14-mc DX.

UPSIDE-DOWN, LOW-ANGLE, 7-14 MC ANTENNAS FOR DX

One of the problems with any of the standard vertical radiators on 14 and 7 mc is that the bulk of the radiation comes from the base of the antenna—the portion closest to the ground. Hence, this signal is the one most apt to be interfered with by neighboring objects. For this reason, designs have been worked out which, in effect, are upside down. The high-current, radiating portion of the antenna is at the top instead of the bottom, as shown in Fig. 9-9.

This type of antenna is fed, of course, at a maximum voltage point, which means that the impedance is far too high for the usual coax feed. Instead, the coax line from the transmitter is link-coupled to a tank circuit which tunes to the operating frequency, and the antenna is fed off the hot end of the tank circuit.

Fig. 9-10 shows dimensions and layouts for antennas of this type for both 14 mc and 7 mc. Both antennas are directional at right angles to the plane of the antenna. What we have, in effect, is a broadside array with the top half of half-wave sections folded down and joined. The gain of this simple array is approximately 4 db. Enough to make a real difference in your output, and, more important, concentrating that output at the extremely low angles so necessary for DX. As shown in Fig. 9-10, the antenna can be constructed entirely of wire; however, the alternate method, combining wire and tubing, is the preferred technique.

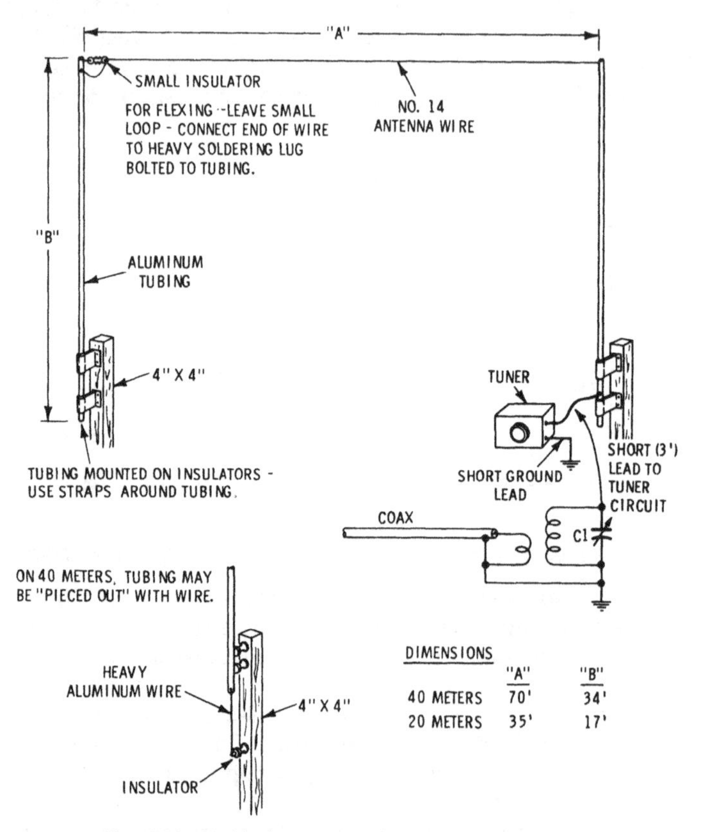

Fig. 9-10. Upside-down vertical construction details.

Antennas of this type are capable of surprisingly good results, considering their simplicity. W8BIE, a Michigan ham who has done considerable test work with this configuration, reported that it consistently out-performed a pair of ground-plane antennas fed in phase. A West-Coast ham developed a somewhat more complex antenna utilizing the same basic idea; it proved outstanding for both 7- and 3.5-mc DX. Fig. 9-11 illustrates the layout.

Tune-Up

Either of the antennas just described tune up in the same fashion:

1. The tank circuit (which should use a low-loss coil and a variable capacitor with the same spacing as the plate-tuning capacitor of the transmitter) is tuned to resonance by coupling a pick-up loop and bulb to it.
2. The coil which connects to the feed line is then adjusted —both coupling and number of turns—to give the lowest

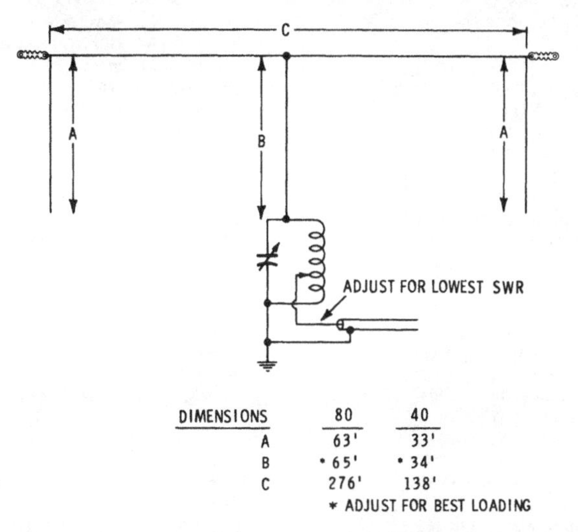

Fig. 9-11. 3-element upside-down vertical for 80- and 40-meter DX.

SWR, as indicated by an SWR meter. The final adjustment is for the setting of C1 for the lowest SWR. Ordinarily, it will be found that the low SWR point can be shifted to the most-used frequency within a band simply by making this adjustment.

Index